LA BOULANGERIE

熱 石 上 的 麵 包

—— GRÉGOIRE MICHAUD ——

推薦序

我深信一個好的糕餅師，應懂得精準計算材料，並確切明白酵母、小麥和糖融合後的化學作用，才能創造美味的糕點。

一個傑出的糕餅師，除了擁有優質糕餅師的精湛技藝外，還要比實際懂得更多。Grégoire 對糕餅食藝熱誠滿腔，有時甚至近似瘋狂。他的創作來自其興趣。勇於嘗試的他，產品常常鋒芒畢露，令人目眩。

沒錯！在這個殷勤又觸覺敏銳的瑞士餅師體內滲着的反叛份子......便是創新意念的動力。

我因服務於非牟利機構「愛心麵包」，有幸認識 Grégoire，他代表酒店捐贈麵包予有需要人士，尤其是受忽視的老人。

Grégoire 是敝機構最早捐贈者之一。每星期最少一次，我會到他的廚房收取麵包，品種繁多，有牛角包、鬆餅、丹麥包、心形包等。無論何時，他會預先籌劃，如遇老人有特別需要，他必定體恤地順應他們的要求。

他不但只把麵包捐給有需要的人，還主動參予到貧窮的地方派發麵包。在某個星期日的下午，我帶 Grégoire 和一群義工到臭名遠播的籠屋，探訪居住在那裏的老人，那地方的居住條件比狗屋還差，我們也到過露宿者之家和屋村獨居長者之家派麵包。當工作完成，我注意到 Grégoire 的耳朵變紅，熱淚盈眶，為受助者的困境深感難過。Grégoire 隨即在其網誌，與大家分享當天的所見所聞。

對的！ Grégoire 的心充滿熱情，這就是他做美妙麵包的秘密材料。

CELENE P. LOO

愛心麵包創辦人
www.givingbread.org

I believe a good pastry chef is one who is disciplined with the exact measurements of the ingredients and understands the chemistry that occurred between the yeast, wheat and sugar etc. with the aim to create the best product.

An OUTSTANDING pastry chef, however, is one who has mastered all the skills of a good chef and more - he has PASSION. Grégoire Michaud is definitely not lacking in passion, which sometimes borderlines madness. Which is why his creation is always interesting, and often courageous with a hint of edginess.

Yes! Beneath the polite sensibilities of this Swiss chef, there is a rebel... with a cause.

I had the privilege of getting to know Grégoire through his contribution in Giving Bread, a non-profit organization which collects fresh and leftover breads from restaurants, hotels and bakeries and give them to the needy especially the neglected elderly.

Grégoire is one of Giving Bread's earliest donors. At least once a week, I visit him in his kitchen to collect the gourmet breads which he donates to Giving Bread - a delicious spread of croissants, muffins, Danish, hearty breads etc. He would plan ahead of time whenever he knew I was coming to collect breads and often accommodate my requests for more supplies when we received special needs from elderly centers.

Once he asked to volunteer in our distribution activity as he was not contented with only baking and giving us the breads. He wanted to go to the slums. On that particular Sunday afternoon, I took Grégoire and a group of volunteers to the infamous cage homes where old men lived in conditions much worse than domestic dogs. We also went to a homeless shelter and an elderly housing estate. When we were finished distributing all the packs of breads, I noticed that Grégoire's ears were crimson and his eyes teary. He was visibly moved by the plight of the needy and sped home to write a heartfelt article about Giving Bread in his blog.

Yes! Grégoire has PASSION. And that's the secret ingredient in his breads which make them so heavenly.

CELENE P. LOO

Founder, Giving Bread
www.givingbread.org

推薦序

Greg:

　　我很榮幸給你的新書寫序。你是烘焙專業界的中堅份子，與這書相比，重要得多了。編寫這書，你把祖先們最寶貴的才幹延續和傳承開去。

　　你我都一樣，在法國、瑞士或任何國家，午夜時間總是留給烘焙師們。當家人和朋友都進入夢鄉，我們卻悄悄地開始工作。在總烘焙師指揮下，各就各位，搓揉麵包機在運作，焗爐在預熱，廚房裏瀰漫着發酵的獨特香味，寓意着我們正式投進這食物藝術。

　　時至今天，這份投入已變為你的部份責任，書中每個食譜給予的提示和意見，足以令同業有所啟發，相信你的目標已達到了。

　　我習慣以雙手配合腦袋去表達我們的專業；沒有用心的思考和自省，造出來的製品實在一無是處。

　　這本書開首數頁，談論麵包製造的理論，這是非常重要的。烘焙在不斷演變和進化，雖然毋須每個細節都作理性分析，但如果這是魔法的原動力，我們必須盡力理解、估計和預先部署。從錯誤中獲得啟蒙，我們的烘焙技術才會有所進步。

　　麵包的價值無分地域，你把麵包帶到香港，繼續獻身於這個行業，並與別人分享你的熱情，讓讀者從中感到製作麵包的樂趣。

　　我向集烘焙師、作者和旅遊家一身的你致意！

Xavier Honorin
Boulangerie 大賽世界冠軍

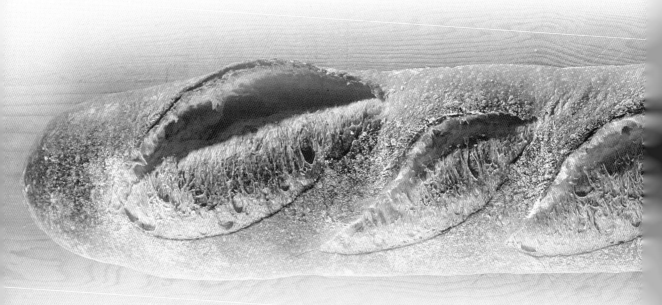

XAVIER'S FOREWORD

Greg,

It is a real pleasure for me to preface your work. So much more than a book, you are placing an important stone in the edifice of our profession. You are leaving your mark, a trail that others can follow. In writing this book, you are continuing the noble tradition of our forefathers by transmitting their precious savoir-faire.

For you as for me, in France, Switzerland or whatever the country, night time is reserved for our bakers. Our families and friends are sleeping, and discretely, our work begins. Our master bakers guide us, the kneading machines are running, the ovens are warming, the leavens fill the kitchen with their particular aroma and our commitment to this art starts to rise.

This commitment today becomes part of you, and if you find that as a result of this book, you inspire other bakers, even just through one recipe, by giving them a hint or a simple idea, you have achieved your goal.

I am often in the habit of presenting our profession as the coordination of the hand and the head.

Nothing worth doing comes without intense reflection.

The first few pages of the book discuss the theoretic approach of our profession. This is very important. Baking is evolving, and even if it is not necessary to look for a rational explanation for everything, if that aspect of magic subsists, we must do our best to understand, anticipate and plan ahead. Our mistakes must bring us enlightenment and the technology of baking is there to help us.

The values of bread are universal. You continue to show from your home in Hong Kong, you're your commitment is intact. The trip you have embarked upon allows you to share you passion with others; it is now up to your readers to discover it.

My compliments to the baker-writer-traveler that you are.

Xavier Honorin
Champion de Monde de Boulangerie

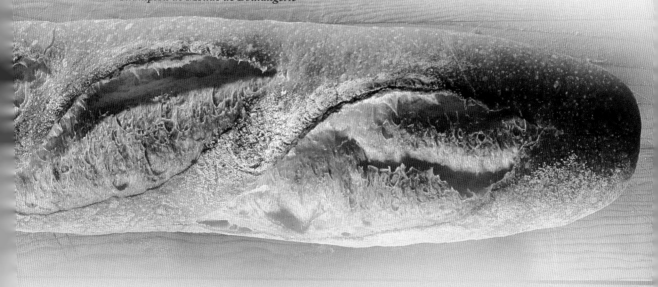

前言

雖然這書介紹的麵包不少與我第一本作品《麵包教室》相同，但在這書裏，我改以家庭式烘焙為主導概念，撤除不便之處以及可能造成困難的做法；我剔走了繁瑣的技巧和複雜的科學理論，讓這本書成為大眾化的食譜，化繁為簡，方便在家製作。我還加入許多創新食譜 —— 所有的新嘗試皆能從你的焗爐烘出來！

這書有很多一流的食譜，特別為新手而設；至於同業，則可從中找到創意，或將優質烘焙方法實踐在專業領域中。

在本書裏，我特別談及在家中如何運用烘焙石以及採用快速乾酵母取代新鮮酵母；因為新鮮酵母不普及，不易在雜貨店找到。除此之外，我又會解釋設有噴射水蒸汽功能的家庭式焗爐。在麵包製作中，麵粉是最重要的成份，為了切合需要，書中提供一個快速指南，以方便在超級市場內找到所需的麵粉。我還在每個食譜附加百分比，讓專業烘焙師可自由調校份量。

對於我而言，在家烘焙美食，招待親朋好友，是生命裏的珍貴時刻。從搓揉過程，獲得的體會，是一份禮物，一段深深印在腦海的美好回憶。一條麵包，最偉大之處是能與人分享，亦是一場歡宴不可缺的美食。當麵包從焗爐取出，傳出隱約的碎裂聲響，陣陣的香氣瀰漫在家裏，魔法便開始了。

優秀的麵包揉合了經典和傳統，亦是任何一位麵包愛好者的終極所愛。於是，麵包一直以來都是主要食糧，千百年來都沒有失去吸引力，慶幸還能保持原汁原味，沒有出現「分子料理法包」或「黑麥包魚子醬」！

今天，真味麵包的價值也許不及從前；感謝在世界各地的積極推動，真味道的麵包終於再度抬頭！其實我能為近代麵包的發展大造文章，不過，借 Robert Orben 以下一句話正好作為總結：

> 「我明白大型食品公司已研發出不刺激眼淚的洋蔥。我想他們必能做到 ——沒有味道的麵包了。」

閔言樂

PREFACE

Largely drifted from my first baking book "Artisan Bread", I have oriented this second edition towards baking your own bread at home with all the advantages, inconvenience and great challenge it induces. I left the complicated techniques and sciences aside for the moment and have kept the popular recipes, modified them for easier home baking. I also added other great new recipes–all tested and baked in our very own ovens!

This book has perfect recipes for the novice home baker; yet, the advanced baker will also find creative ideas and great quality baking method to be used in a professional environment.

In this second edition, I thought helpful to talk about the usage of baking stone at home and as well, I have adapted all the recipes using dry yeast instead of fresh yeast, the latter not being the most common in today's grocery stores. The exhaust is also a feature that is rarely found on home ovens, thus I am also explaining that concept. Flour is a typical topic of uncertainty in home baking, and you'll find an easy guide to which supermarket flour is suitable and for what purpose. Additionally, for you, baking experts out there, I have added the baker's percentage in each recipe.

To me, baking at home for family and friends is one of life's most rewarding moments. From scratch, you create an experience, a gift, a memory that will be etched in minds as a fond time. One of bread loaf's greatest features beside of being an amazing food is that it is shared amongst people; bread is convivial. When your bread comes out of the oven and starts its subtle crackling noise, dispersing its fairy tale scents all around home, the magic happens.

Good bread in its classic and traditional form is the ultimate goal of any bread lover; with this, bread is a staple food that crossed centuries without loosing any of its appeal–almost untouchable. For that matter, there was never any 'espuma of baguette' or 'rye bread caviar' bred from modern cuisine; and I'm thankful for that.

Today, the values of real bread in our society aren't generally as deeply anchored as in the past; yet, thanks to a solid worldwide movement, we see real bread raising and rising again! I could have written a whole chapter on recent bread development, but I thought Robert Orben summed it all pretty well when he said:

"I understand the big food companies are developing a tearless onion. I think they can do it - after all, they've already given us tasteless bread."

Grégoire Michaud

目錄 CONTENTS

2 　推薦序 CELENE'S FOREWORD

4 　推薦序 XAVIER'S FOREWORD

6 　前言 PREFACE

11 　一粒麥子的告白 CONFESSION OF THE WHEAT

16 　麵包是怎樣做出來？ HOW DOES IT WORK?

44 　閱言樂的麵包廚房 ARTISAN BREAD AT GRÉGOIRE'S KITCHEN

甜包
A TOUCH OF SWEETNESS

48 　藏紅花香草麵包
　　SAFFRON VANILLA CUCHAULE

50 　乾果長麵包
　　THICK DRIED FRUIT LOAF

53 　西梅核桃麵包
　　PRUNE WALNUT BREAD

56 　葡萄乾碧根果仁麵包
　　PECAN AND RASIN BOULE

58 　黑朱古力酸櫻桃麵包
　　DARK CHOCOLATE & SOUR CHERRY BREAD

60 　國王麵包
　　KING'S BREAD

62 　椰糖牛油粒圓包
　　COCONUT PALM SUGAR ROLL

64 　蜜餞薑粒皇家麵包
　　CANDIED GINGER PAN D'ORO

67 　啤梨杏仁牛油撻
　　PEAR FRANGIPANE BRIOCHE

69 　榛子朱古脆卷
　　HAZELNUT PRALINE FLAKY ROLLS

72 　珍珠糖及朱古力忌廉麵包
　　PEARL SUGAR & CHOCOLATE BREAD

75 　甘薯琥珀核桃包
　　SWEET POTATO & CARAMELISED CHINESE WALNUT "COCOTTE"

鹹包
THE ARROGANCE OF SALT

80　德國番茄辣椒蝴蝶包
　　TOMATO CHILI PRETZEL

82　辣椒格魯耶芝士條
　　GRUYERE CHEESE & PAPRIKA TWIST

84　黑胡椒茴香脆條
　　BLACK PEPPER & FENNEL TARALLI

87　法國紅辣椒薄脆
　　PIMENT D'ESPELETTE LAVOSH

90　蒜頭香芹牛油軟包
　　GARLIC PARSLEY BRIOCHE NEEDLE

92　煙肉香草包
　　BACON & HERBS EPI

94　蒜蓉香草俄羅斯軟包
　　RUSSIAN PAMPUSHKA

96　新鮮迷迭香方包
　　FRESH ROSEMARY PAVÉ

98　小洋蔥細葉芹小圓包
　　BABY ONION & CHERVIL BITES

101　帕爾馬火腿芝麻菜比薩
　　ARUGULA & PARMA HAM PIZZA

104　小薯仔檸檬百里香意大利香草包
　　GRENAILLE POTATO & LEMON THYME FOCCACIA

106　普羅旺斯法式香草扁麵包
　　PROVENCAL FOUGASSE

108　番茄乾意式煙肉包
　　SUN DRIED TOMATOES & PANCETTA ROLLS

110　希臘黑橄欖法式麵包
　　KALAMATA OLIVES BAGUTTE

112　大蘑菇鄉村麵包
　　PORTOBELLO MUSHROOM COUNTRY BREAD

114　蘋果酒浸葡萄乾芫荽煙燻煙肉包
　　CIDER RAISINS, SMOKED BACON & CORIANDER ROLLS

117　紅甜椒瑞士芝士卷
　　RED BELL PEPPER & BAGNES CHEESE ROLLS

120　油浸蒜頭牛至扁麵包
　　CONFIT GARLIC & OREGANO SCHIACCIATA

122　香脆松子羅勒卷
　　CRISPY PINE NUTS BASIL ROLLS

巧味包
ALMOST UNTOUCHABLE

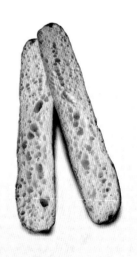

126 全麥愛爾蘭梳打麵包
WHOLE WHEAT IRISH SODA BREAD

128 種籽麵包
SEEDED OVAL

131 蕎麥枕頭麵包
BUCKWHEAT PILLOW BREAD

134 洛神花黑麥麵包
LIGHT ROSELLE RYE BOULE

136 孜然波爾多皇冠包
CARAWAY BORDELAISE CROWN

138 法式長麵包
FRENCH BAGUETTE

140 鄉村麵包棍
FABRICE COUNTRY STICK

142 斯佩爾特小麥農夫麵包
SPELT FARMER BREAD

145 全麥吐司
WHOLE WHEAT PAIN DE MIE

148 海藻檸檬絲吐司
SEAWEED LEMON PAIN DE MIE

150 經典意大利軟包
CLASSIC CIABATTA

152 瑞士辮子麵包
PETITE TRESSE AU BEURRE

154 字母包
CRUMB

156 辣根亞麻籽粗麥方塊包
HORSERADISH FLAXSEED SEMOLINA SQUARES

158 烘焙術語詞彙 GLOSSARY

160 鳴謝 ACKNOWLEDGEMENT

一粒麥子的告白
CONFESSION OF THE WHEAT

小麥 THE WHEAT

在植物學來説，有30,000多個品種的小麥。麥可分為20多類，在眾多小麥裏，斯佩耳特小麥和硬粒小麥最是廣為人知，然而一般小麥，粗略計算也有超過200種主要小麥類適合用於生產麵包粉。

Botanically, there are more than 30,000 varieties of wheat. It is a grass that is cultivated in most countries around the world. Wheat (Triticum) falls into more than 20 categories; amongst others are the Spelt and the Durum that are widely known, but Common Wheat, that counts more than 200 species is the main kind of wheat used for producing bread flour.

胚乳 ENDOSPERM

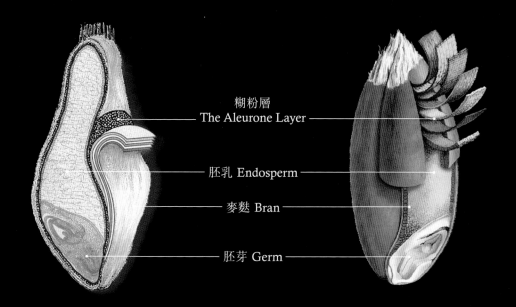

糊粉層
The Aleurone Layer

胚乳 Endosperm

麥麩 Bran

胚芽 Germ

小麥粒結構圖 Composition of a Wheat Kernel

胚乳佔全粒麥粒總重量的83%，亦是白麵粉的澱粉質來源。一顆麥粒包含20,000白麵粉微粒，胚乳佔全顆麥粒最大部份的蛋白質。

It represents about 83% of the kernel weight. It is the source of white flour due to the starch; there are about 20,000 particles of white flours in one grain. The endosperm contains the greatest share of the protein in the whole kernel.

麥麩 BRAN

麥麩佔全顆麥粒重量約14.5%，全麥粉會包含麥麩，它也可獨立使用。麥麩是微粒粉層磨成粉末後的主要副產品。

The bran is about 14.5% of the kernel weight. Bran is included in whole wheat flour and is also available separately. It is the protective layer of the kernel and the main by-product from flour milling.

胚芽 GERM

胚芽約佔麥粒重量的2.5%，胚芽是種籽的可發芽部份。由於它的脂肪會限制了貯藏期，讓麵粉變質，一般會被分隔。只有全麥粉才包含胚芽。

The germ is about 2.5% of the kernel weight. The germ is the embryo or sprouting section of the seed, usually separated because of the fat that limits the keeping of quality flour. Wheat germ is available separately and is included in whole wheat flour.

糊粉層 THE ALEURONE LAYER

糊粉層是圍繞種籽的胚乳的單一細胞層，性質柔軟。在含有澱粉質胚乳的穀物例如小麥或裸麥，糊粉層約佔穀粒的30%蛋白質。

The aleurone layer is a single cell layer that surrounds the endosperm tissue of the seeds. In cereals with starchy endosperm such as wheat or rye, the aleurone layer contains about 30% of the kernel's proteins.

蛋白質 THE PROTEINS

小麥穀粒有兩組蛋白質，水溶性的球蛋白和白蛋白是人類從糧食中可獲取的重要蛋白質。

另一組是非水溶性的麥穀蛋白和麥醇溶蛋白，它們一起形成麵筋。麵筋在製造麵包糰中扮演重要角色。麵筋使搓揉好的麵糰有彈性，容許麵糰發酵並令烘焙製品產生各異其趣的質感。麵筋的彈性與麵粉所含麥穀蛋白成正比。

對麵筋敏感的人經常會出現腸胃疾病，它是一種對麥醇溶蛋白消化分解物的不正常免疫反應。有這種毛病的人愈來愈多，因此，建議在麵包作坊裏同時供應一些無麩質的麵包。

There are two main groups of proteins in the wheat grain. The water soluble proteins are globulin and albumin precious proteins in the human diet.

The other proteins are non-soluble in water and they are the glutenin and gliadin that together form the gluten. The gluten plays a key role in the making of bread dough. Gluten gives kneaded dough its elasticity, allowing leavening and contributing to form a certain texture in baked products. The elasticity of gluten is proportional to its content of glutenins.

Celiac disease is the principal disorder caused by gluten sensitivity. It is an abnormal immune reaction to digestive breakdown products of gliadin. The population suffering from such disorder is increasing and it is therefore recommended to offer gluten-free bread in an Artisan bakery.

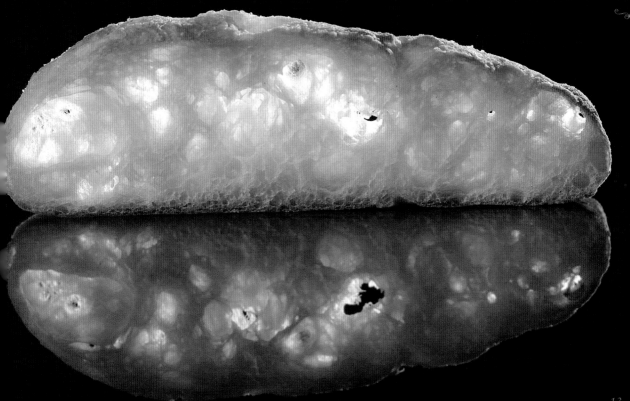

不同種類的小麥 DIFFERENT KINDS OF WHEAT

用以造麵粉的小麥，主要來自美國、加拿大和澳洲，其他國家均有出產小麥，不過主要生產國均來自前述的北美和澳洲等地。一般常用的小麥種植種類如下：

Mainly, wheat used for bread flour originates from the USA, Canada and Australia. Other countries are also producing wheat, but in general, they also import wheat from these countries. The most popular kinds of wheat grown for the production of flour are:

冬天硬紅麥 HARD RED WINTER

優質蛋白質層，適合研磨和能配合烘焙特質。一般用作生產發酵麵包。

Fairly good protein levels; good milling and baking characteristics. Used to produce leavened bread.

春天硬紅麥 HARD RED SPRING

含有豐富蛋白質，這種極品小麥經細心研磨成麵粉，可以製造出極品麵包。

Contains the highest percentage of protein, making it an excellent wheat for bread flour with superior milling and baking characteristics.

冬天軟紅麥 SOFT RED WINTER

高收成，但是蛋白質含量不高，可造餅粉。

High yielding, but moderately low in protein. Used for cake flour.

硬粒小麥 DURUM

一種很硬的小麥，主要用作製作粗麵粉以生產意大利式粉麵類製品。

One of the hardest kind of wheat; mainly used to make semolina flour for pasta production.

硬白小麥 HARD WHITE WHEAT

與紅小麥特質差不多（除了顏色基因），這款小麥柔軟、含甜味，纖維質平均等，研磨及烘焙特質相類似。

Closely related to red wheat (except for color genes), this wheat has a milder, sweeter flavor, equal fiber and similar milling and baking properties.

軟白小麥 SOFT WHITE WHEAT

它的用途與冬天軟紅麥類同（除了造麵包，也適合一般烘焙製品）。蛋白質含量低，但是屬高收成的小麥。

Used in the same way as Soft Red Winter (for bakery products other than bread). Low protein, but high yielding.

麵包是怎樣做出來？
HOW DOES IT WORK?

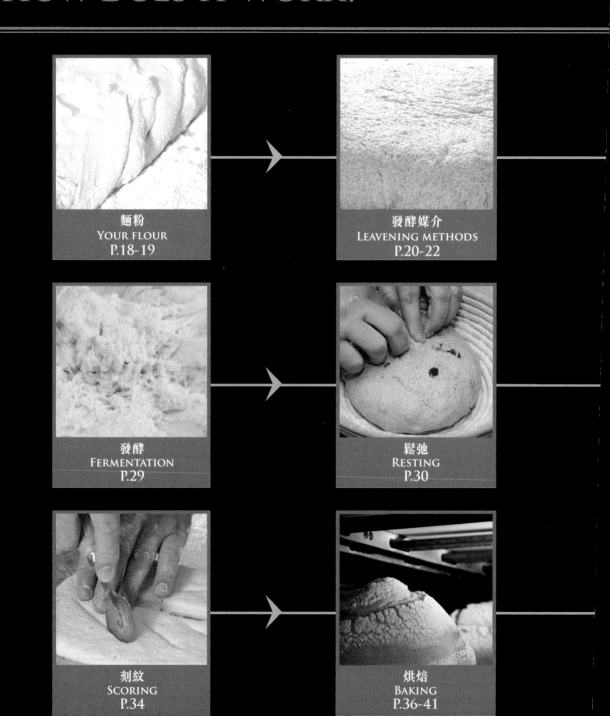

麵粉
YOUR FLOUR
P.18-19

發酵媒介
LEAVENING METHODS
P.20-22

發酵
FERMENTATION
P.29

鬆弛
RESTING
P.30

刻紋
SCORING
P.34

烘焙
BAKING
P.36-41

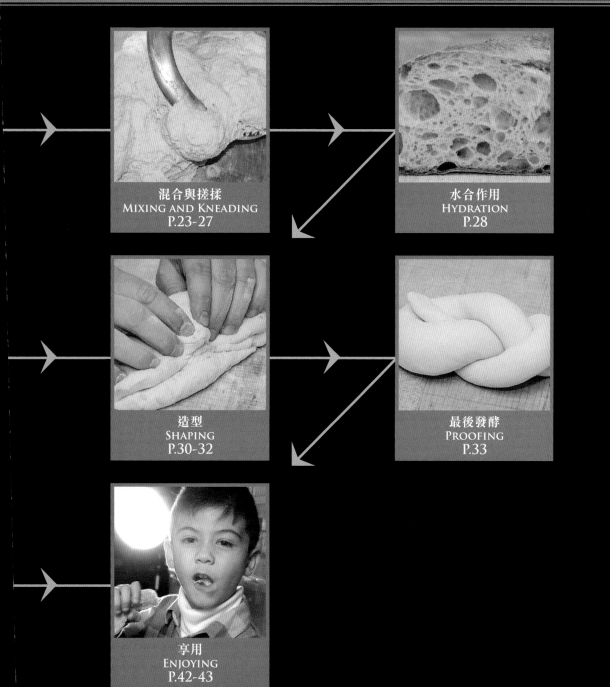

混合與搓揉
MIXING AND KNEADING
P.23-27

水合作用
HYDRATION
P.28

造型
SHAPING
P.30-32

最後發酵
PROOFING
P.33

享用
ENJOYING
P.42-43

麵粉 / YOUR FLOUR

踏進超級市場，有多種多樣的麵粉，名字和品牌各有不同，要找適合做麵包的麵粉真傷腦筋，下頁的速查表，能幫助你找到最適合書中食譜的麵粉。值得一提，麵粉的麥麩質含量與蛋白質沒有關係，例如典型的麵條麵粉（型號是00）含非常豐富的蛋白質。此外，超級市場的全麥麵粉和裸麥粉就沒有劃分太多磨製成品類別，反倒容易挑選。（見表格1）

Going to the supermarket and hunting for suitable flour for baking bread can be a real headache with so many different names and different brands. The quick reference table below will help you find the most suitable flour for the recipes in this book. It is worth noting that the ash content is not related to the protein level, meaning for example, typical pasta flour (type 00) can be very strong in protein. Additionally, supermarket's whole wheat flour and rye flour are not divided in so many kinds of milling and thus easier to select. (See table 1)

書中 IN THIS BOOK	常用名字 COMMON NAME	特徵 / 用途 FEATURE / USAGE
45 號麵粉 Flour type 45	純白麵粉 Plain white flour	低蛋白物質，適合做蛋糕、曲奇和軟麵包 Lower protein content, cake, cookies and soft breads
55 號麵粉 Flour type 55	在食譜內，可替代 45 和 65 號麵粉 In recipes, can be used for both type 45 or 65	
65 號麵粉 Flour type 65	強力白麵粉 Strong white flour	高蛋白質，適合做外層香脆、強筋線的麵包 Higher protein, crusty bread, stronger gluten thread
00 號麵粉 Flour type 00	麵粉型號 00 Flour type 00	非常精緻麵粉，蛋白質良好，適合做麵條、薄餅 Very fine flour, good protein, pasta, pizza

表格 1　Table 1

小麥麵粉的類型由麥麩質含量或被抽出來的可用成份而決定。以每100公斤穀粒計算，可被抽取的麵粉量，所得份量稱為磨製粉的生產量。麥麩含量從烘焙麵粉經用900℃烘焗2小時，餘下的礦物質所計算。透過未被燃燒的礦物質，磨坊可計算出麥麩質含量。由於不同國家的計算方法各異，大部份國家均有指定法律管制不同種類的小麥麵粉。（見表格2）

The type of wheat flour is mainly defined by its ash content or by its extraction level. The extraction level is the quantity of flour obtained after the milling of 100 kilos of clean grain. This is also called the milling yield. The ash content is measured by the minerals left after baking flour at 900°C for about 2 hours. Minerals won't be combusted; thus, the miller can measure the ash content. Different countries use diverse ways of measurement for wheat flour. Most countries have specific law governing specification of flours and their designation. (See table 2)

灰質 ASH	蛋白質 PROTEIN	麵粉類別 FLOUR TYPE		
		美國 U.S.A	法國 France	歐洲 Europe
~0.4%	~9%	糕餅麵粉 pastry flour	45	450
~0.55%	~11%	多用途麵粉 all-purpose flour	55	550
~0.8%	~14%	高筋麵粉 high gluten flour	80	800
~1%	~15%	純質麵粉 first clear flour	110	1100
>1.5	~13%	白全麥麵粉 white whole wheat	150	1500

表格2　Table 2

根據表格2顯示，得知美國、法國和歐洲麥粉的差異性，例如：美國依據麥麩質和蛋白比例劃分麵粉類別，歐洲國家與美國相若，以每100克為單位，粗略計算乾燥物，至於法國則以10克以上計算。你會發現書中食譜有多種麵粉，最為常用的只有3種，這就是白小麥麵粉、全麥麵粉和黑麥麵粉。

In the table 2 we can observe the difference between different regions. For example the US uses the ash and protein content to define the type of flour; European countries are roughly the same calculating on 100 grams of dry matters while French are calculating it over 10 grams. You will discover many types of flour along the recipes of this book; however there are 3 main types widely used in baking. This is the white wheat flour, the whole wheat flour and the rye flour.

發酵媒介

乾酵母 / 乾依士

這裏有兩種主要乾酵母：活性乾酵母和即用乾酵母。活性乾酵母必須溶於38℃的溫水，停放15分鐘後，效果理想。如果水的溫度較低，例如水溫只有20℃，其發酵能力會降至約35%，即用乾酵母在超級市場內最常找到。它在物料融合前便要混和，確保所有酵母小粒子正常溶化。混合時間太短，會使部份粒子殘留而未能溶化。書中食譜採用這款容易使用的快速乾酵母，方便家庭式烘焙之用。

新鮮酵母與乾酵母的配份比例，按照換算準則是每10克乾酵母相等於35克新鮮酵母。乾酵母是新鮮酵母脫水後的製品，主要分別在於從新鮮酵母抽出的水份。就發酵效果而言，它們是一樣的，但所謂完美主義的烘焙師或許會說使用新鮮酵母更能帶出其芬芳香味。

LEAVENING METHODS

DRY YEAST

There are two main kind of dry yeast. They are the active dry yeast and the instant dry yeast. The active dry yeast must be dissolved in water at 38°C and rest for a period of 15 minutes in order to have efficient results. If the temperature is much lower, for example at 20°C, the fermenting power will be reduced of about 35%.

The instant dry yeast is the most popular found in supermarket. It is important to mix the instant dry yeast at the beginning of the mixing in order to have all the granules of yeast dissolved properly. A too short time of mixing could leave some granule not dissolved. The recipes of this book are using instant dry yeast making the recipes home-baking friendly.

In terms of equivalence between the fresh yeast and dry yeast, the general rule is that 10 gm of dry yeast is equal to about 35 gm of fresh yeast. Dry yeast is basically dehydrated fresh yeast; therefore, the main difference is the extraction of water from the fresh yeast. In terms of fermenting effect, they are equal. But the so called 'purist' bakers might say that fresh yeast gives better aromas.

新鮮酵母

　　新鮮烘焙酵母是5微米單細胞培植菌。它能製造出發酵物質如糖、果汁或葡萄乾。當與氧氣相遇，酵母會繁殖，例如在麵糰裏，酵母發酵會產生二氧化碳，導致麵糰脹大，酵母的生物名稱為「Saccharomyces cerevisiae」。在自然界有許多野生的酵母，當我們製造麵種時，就要用上野生的酵母了。

　　酵母對不同熱度很敏感，並於溫度約50℃內會死掉。它由75%水份、12%蛋白質和13%碳水化合物（即醣）組成。新鮮酵母能被壓成細小塊或可供大量購買。在溫度約5℃，可保存2~3星期。就物流或倉存控制為目的，可置冰庫貯藏約6個月，但其效力會續漸降低。值得一提，酵母要慢慢解凍，完成後就必須立即使用，因為解凍後的酵母會化成液體。

FRESH YEAST

Fresh baker's yeast is a cultivated unicellular mushroom measuring about 5 μm. Yeast is creating the fermentation of substances like sugar, fruit juices or raisins. With oxygen, yeast is multiplying, for example in dough; yeast ferments and produces carbon dioxide (CO_2), causing the dough to rise. The biological name of yeast is: Saccharomyces cerevisiae. There are many kind of wild yeast in nature and when we make a sourdough, for example, we are using these wild yeasts.

Yeast dies at around 50°C and is very sensitive to thermal differences. Yeast is composed of water (75%), proteins (12%) and carbohydrates (13%). Fresh yeast can be found pressed in small blocks or can be purchased in bulk. It can be kept for about 2 to 3 weeks at 5°C, and for logistic or stock control purpose, it can be frozen to up to 6 months; however, the effect of the yeast will decrease. It is important to defrost the yeast slowly and to use it immediately once thaw, also, after thawing, the yeast may become liquefied.

速成麵種

　　速成麵種首要則規是用同等份量的麵粉和清水，再混合少許酵母，放進膠盒作長時間發酵。在冰箱的發酵時間24小時，而置於室溫則只需數小時。當所有糖份已進入麵糰協助發酵，麵種將開始喪失能力。

　　速成麵種完成時，確保麵粉完全混合而沒有粉塊。切記不要在麵糰放鹽，它會令所有材料的發酵力降低。短時間發酵麵糰能帶出芬芳香味，長時間留在麵包裏。同時麵包層的顏色呈奶白。

POOLISH

The rule of thumb for poolish (or sponge) is an equal part of flour and water mixed with a little yeast. It will proof over a relatively long period of time in a plastic container. The fermentation time can be up to 24 hours in a fridge or fewer hours at room temperature. The poolish will start loosing its power once all the sugars present in the dough will be consumed by ferment.

While making poolish, ensure the flour is well mixed to avoid lumps of flour. Also, it is important not to add salt in a poolish as it will reduce greatly the fermentation all together. The polish brings aromas and a longer period of conservation to the bread. As well, the crumb color will be cream rather than white.

酸酵麵種

　　真正麵包由麵種為開端——這是完美主義的烘焙師說的。麵種是發酵麵包的最典型和天然方法，它能令麵包外層特別香脆和產生討喜的麵包屑。相對於傳統的發酵方法，麵包的保鮮期較長。這種方法能使麵包與天然細菌隔離，麵種的微酸味道可平衡麵包的味道，每個烘焙師會因應製品需求而發展其秘方，一個好的麵種能不斷更新和保存多年。

SOURDOUGH

Real bread starts with sourdough - will say the purist bakers. Sourdough is the exemplary natural method for leavening bread. Sourdough gives to the bread a spectacular crust as well as a very pleasant crumb. Breads will stay fresh for a longer period compared to conventional leavening methods. This method also prevents bread stalling with its natural bacteria. The slight acidic flavor of the sourdough gives the bread a great balance in taste. Each baker develops its own method and recipe according to the needs of the production. A good sourdough can be refreshed and maintained over many years.

混合與搓揉

混合麵糰前需要計算和量好各材料，這是非常重要。使用慢速，徹底攪拌材料，以求各材料正常融合，這是通質化過程。當速度增加，在機械運作下，麥穀蛋白和醇溶性蛋白會構成麵筋網絡。澱粉細胞摩擦時，溫度升高，容許吸入更多濕度。經研究和分析這特別的摩擦過程——麵糰流變學，它是麵包製造的要素。當搓揉麵包的速度增加，便不會有水或麵粉的痕跡可尋。烘焙師在此時應該觸摸麵糰，感受它是否太軟或太硬，作出調節。任何麵粉或清水添加於麵糰內，就要早些調足所需，好讓澱粉有足夠時間轉化。

自動溶解法是指開始搓揉麵糰時，一個慣常用的程序。此過程可應用於許多食譜中要把清水和麵粉混合，讓其靜止20分鐘至1小時，待麵粉脫水發酵，正如蛋白酵素類物質會開始破壞麵粉的蛋白質。自動溶解法會令麵糰強壯和易於擴張，因為澱粉質早點脫水，麵糰的伸展屬性會比較好。當麥穀蛋白和醇溶性蛋白開始被破壞，麵糰的擴展度會增加。這過程需要在短時間內把兩種材料混合，不能作長時間攪拌。

MIXING AND KNEADING

It is important to measure and weigh each ingredient before mixing any dough. The mixing of ingredients must be done in slow speed in order for all the components to blend properly; this is the homogenization process. When the speed is increased, under the mechanical action, the glutenin and gliadin will form a network of gluten. The friction of starch cells will allow them to absorb more humidity as the temperature increases. This particular friction process is being studied and analyzed as dough rheology, an essential factor in bread making. When increasing kneading speed, there should be no more traces of water or flour. The baker should now touch the dough to feel if the dough is too soft or too hard and make any adjustment if needed. If any flour or water is added, it is important to adjust it early enough to allow the starches to be transformed adequately.

A common process to start kneading bread dough is called autolyse. It's a process you can apply to many recipes by only mixing the water and the flour of the recipe and let it rest between 20 minutes to one hour. It will allow the flour to start hydrating and the enzymes, such as protease will start to break down proteins in the flour. The autolyse will make the dough stronger and extensible. Because of the early hydration of starch, the dough will have better stretching properties. While the glutenin and gliadin starts to be broken down, the extensibility of the dough will increase. This process needs to be done just by mixing both ingredients shortly, without stirring for a long time.

最後搓揉過程才添加鹽，可徹底地架起麵筋網絡，且它與吸濕物質可拉緊麵糰。如果必需加入牛油，建議在最後混合時才添加，令麵筋網絡完全建立。牛油會引起澱粉細胞和夾層分離，將會降低麵筋的形成。待最後搓揉之前，烘焙師應該用手把小部份麵糰拉扯檢查它的彈性。正常的麵筋，可看到網絡呈透明而不破損。如果它破裂，就要繼續搓揉。由於小麥的磨研粗糙，其彈性質地會較弱。

The salt is generally added at the end of kneading in order to allow the gluten network to fully form. Salt will tighten the dough with its hygroscopic properties. If butter is needed, it has to be added at the end of mixing to allow the gluten network to form completely. Butter will create an insulating layer between the cells of starch and will decrease the formation of gluten. Before the end of kneading, the baker should check the elasticity of the dough by pulling a small part of dough between his hands. For proper gluten, you should be able to almost see through the dough without breaking it. If it breaks, it needs more kneading. Dough with whole meal flours will have less elastic textures due to the coarse milling.

麵粉裏，其吸濕程度取決於不同因素（見表格3）。

In flour, the absorption level of humidity is determined by different factors: (See table 3)

麵粉的濕潤度 **Moisture content**	低濕潤度 = 高吸濕能力 **lower moisture = higher absorption**
蛋白成份 **Protein content**	低蛋白質 = 低吸濕能力 **lower protein = lower absorption**
麵粉等級、灰質或麥糠成份 **Flour grade, ash or bran content**	高價值 = 高吸濕能力 **higher value = higher absorption**
水溶性蛋白 **Water soluble protein**	高數值 = 高吸濕能力 **higher value = higher absorption**
破損澱粉 **Damaged starch**	高數值 = 高吸濕能力 **higher value = higher absorption**
酶活性（澱粉酵素） **Enzymes activity (amylase)**	活性越高，吸濕能力越低 **greater activity gives lower absorption**

表格 3　Table 3

麵糰搓揉理想是有極限的。烘焙師應在搓揉極限前，停止搓揉。如果超越了極限，會令麵糰過份混合。拉開時會被撕成小塊；又因它含有雞蛋、糖和牛油，將會創造長纖維，在這情況下，我們稱為麵糰

Bread dough has a peak point of optimal kneading. The baker always stops the kneading before the peak point of the extensibility of the dough. If it passes the peak point it will decrease and the dough will be over kneaded. When trying to pull the dough, it would tear into small parts or if the dough contains eggs, sugar and

「燒焦/過度搓揉」。

書中食譜主要用螺旋形混合器或單臂式混合器。少量麵糰可用星球形混合器，為家庭式常用混合器。螺旋形混合器的碗與其臂的移動，背道而馳，意即其臂運行而拂碗固定，增加摩擦力。烘焙師可根據搓揉的運作情況，自行調節，因不同型號的混合器、麵粉、天氣和溫度會帶來變數。

麵包店常用混合器（見表格4）。

butter, it will create very long strings. In this case, we call the dough 'burnt'.

The recipes in this book primarily use spiral mixer and single arm mixer. For smaller dough we also used planetary mixer. A planetary mixer is equivalent to most mixers present at home. The bowl of the spiral mixer moves on the opposite way of the arm movement to increase friction, while in the planetary, the bowl is fixed. The baker himself will judge if the kneading is proper or not. It will vary according to the type of mixer, flour, weather and temperatures.

The following mixers are the most common in small to medium size bakeries. (See table 4)

混合器 MIXER	每分鐘轉數 RPM	類別 CLASS	優點 ADVANTAGE	缺點 DISADVANTAGE
單臂式 Single Arm	28 to 60	傳統式 Conventional	搓揉能力良好 Good kneading	搓揉時間長 Kneading takes long time
雙臂式 Twin Arm	28 to 60	傳統式 Conventional	氧氣容量高 Higher volume of oxygen	麵粉塵較多 Strong flour dust
球式／軌道式 Planetary	可調校 Adjustable	混合式 Mixed	變速能力佳 Great flexibility of speed	尺碼有限 Limited sizes
螺旋式 Spiral	80 to 120	密集式 Intensive	搓揉極佳過程 Excellent kneading process	有過度混合的危機 Higher risk of over-mixing
分叉式 Fork	80 to 120	密集式 Intensive	氧氣容量高 Higher volume of oxygen	有過度混合的危機 Higher risk of over-mixing
雙筒式 Double Cone	80 to 140	密集式 Intensive	能細緻處理麵糰 Dough is handled carefully	只能處理多份量麵糰 Only for large amount of dough

表格4　Table 4

基本溫度借用算術程式來計算出書中食譜的搓揉時間。為了達到計算目的，當混合麵糰時，便需要計算出水溫（見表格5）。

The basic temperature, present in all recipes where the dough is kneaded, follows a simple calculation method. It has for purpose to calculate the temperature of the water needed. (See table 5)

PROCESS OF KNEADING DOUGH
搓揉麵糰的過程

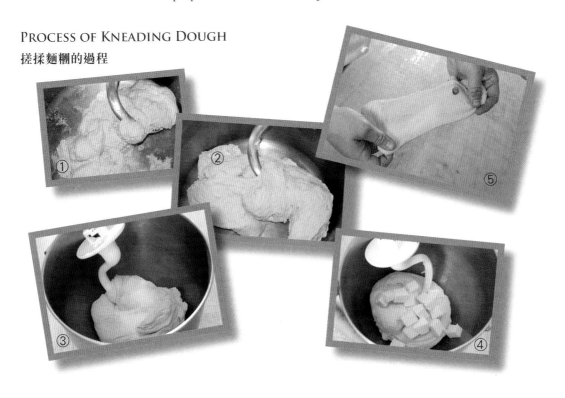

麵糰的目標溫度 Dough temperature target	27 °C
– 麵糰摩擦熱度 Dough friction heating	6 °C
= 目標 Target	21 °C
倍增 Multiplier 3 × 21 °C =	63 °C
室內溫度 Room temperature	25 °C
– 麵粉溫度 Flour temperature	24 °C
= 水溫度 Water temperature	14 °C

表格5 Table 5

麵糰的溫度

在烘焙時，麵糰溫度是非常重要。它決定了靜止和發酵時間，同時烘焙師會因應生產計劃和發酵方法，調節麵糰溫度。一般規則，冷凍麵糰的搓揉溫度由18℃至22℃；而和暖麵糰的搓揉溫度是26℃至30℃。高溫會令澱粉微粒子增加膨脹能力，特別是全麥麵粉。極度和暖的麵糰會膨脹很快，味道和所需的品質未能正常發展。相反地，極冷凍的麵糰就不膨脹和未能發展到所需標準。

有些烘焙工場，備有冷凍水機器給烘焙師供應穩定冷水。然而，在和暖的氣候，沒有機器的協助，很難獲得冷水，為了解決這問題，建議使用冰粒或冰片調節水溫，假設冰粒溫度為-0.5℃，簡單計算方法（見表格5）。

TEMPERATURES OF DOUGH

Temperature of dough is very important in baking. It determines the length of resting and proofing time. Also, the baker will adjust the temperature of the dough according to its production plan and leavening method. As a general rule, the range of temperature for a cold dough kneading is between 18°C and 22°C and for warm dough kneading between 26°C and 30°C. Higher temperature increases the swelling ability of the starch molecules, especially for whole meal flours. Excessive warm dough will proof too fast and not meet the proper development of flavors or required qualities; on the opposite, very cold dough will not proof and not develop to the necessary standards.

In some bakeries, there is a cooling water machine that helps the baker to have a steady supply of cooled water. However, in warm climate, cold water is difficult to obtain without machine. Therefore, the use of ice cubes or ice chips is recommended following a simple calculation assuming that the temperature of the ice is of -0.5°C. (See table 5)

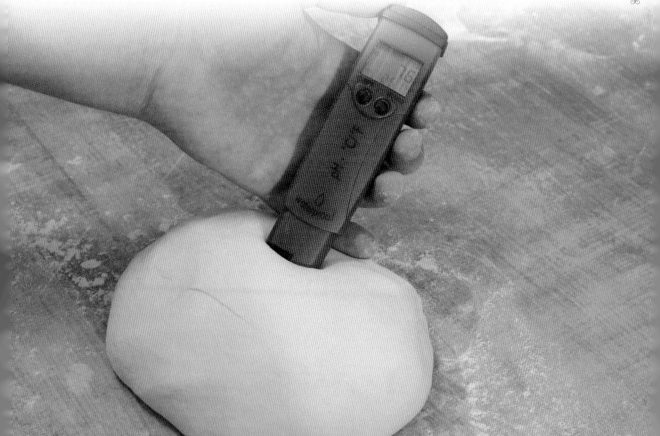

水合作用

　　麵包內碎屑的架構為水合作用層。這水合作用層是食譜中的用量水和產品的其他用作製造麵包的材料相若，水合作用的百分率以麵粉總重計算出來。麵包可達50%至90%不等。有些特別情況會出現較高水合作用層或較低的水合作用層。主導麵糰的水合作用的指標，而牛奶或雞蛋更有助其發展。

　　水合作用過程從混合開始，在均化時期和麵糰進一步發展的變化，已於搓揉麵糰的章節解釋了。

　　根據不同水合作用層，可看到麵包內部碎屑及不同程度的質感在演變中。（見表格6）

HYDRATION

　　The structure of the crumb, the inner part of your bread, is defined by the level of hydration. The hydration level is the amount of water used in your recipe, and as with other ingredients in your bread, the hydration ratio is calculated as a percentage to the total weight of the flour. In typical bread it ranges from 50% to 90%. Some specialties will reach higher hydration level and some even lower levels. While water is the lead indicator of hydration in dough, milk or eggs are also contributing to the hydration level in dough.

　　The hydration process starts at mixing, during the homogenization stage and further continues when the kneading starts as explained in the mixing and kneading chapter.

　　Below are the crumbs and their different textures according to hydration levels. (See table 6)

不同程度的水合作用造成的麵包內碎屑架構
LEVELS OF HYDRATION

表格6　Table 6

製作麵包時，發酵過程的目的是發展麵糰質地和改變其食味。影響發酵的因素有麵粉本身的韌度、酵素（即酶）活動、發酵方法和要求製品的結果。

在發酵時期，酵母會由醣轉化為二氧化碳和乙醇，而二氧化碳會釋出氣體和令麵糰膨脹和伸展，繼而產生乳酸和醋酸。在最初發酵的25分鐘裏，酸鹼值一般會從6.0降至5.0。

任何在麵糰的油脂會幫助氣泡網絡產生額外層，並在受控的發酵溫度下保留令油脂達熔點，例如牛油的熔點大概是33℃。

初次發酵／初步發酵

麵糰經混合和搓揉後，會在混合碗內或已蓋保鮮紙的膠器皿進初次發酵。發酵時間因應個別食譜和方法而變更，時間越久越好，除了某些特定麵包需要預先造型。在初步發酵過程中，麵糰產生二氧化碳，味道和質感便在這時形成。

The goal of the fermentation process in bread making is to develop the dough texture and to influence its taste. Factors affecting the fermentation are the flour strength, enzymatic activity of the flour, the leavening method used and the desired final product.

During the fermentation, the yeast converts the fermentable sugars into CO_2 and ethanol. The CO_2 is then released into gas cells and the dough expands. Lactic acid and acetic acids are generated. Within the first 25 minutes of the fermentation, the pH typically drops from 6.0 to 5.0.

Any fat contained in dough will help creating an extra layer in the bubble network and will help to retain the fermentation gases under a controlled proofing temperature below the melting point of the fat used, for instance butter is about 33°C.

BULK FERMENTATION

After the mixing and kneading of the dough, the bulk fermentation happens either in the mixing bowl itself or in a plastic container covered with a plastic sheet. Time of bulk fermentation varies according to each recipe and method, but the longer the better, except for some specialty bread requiring an earlier shaping. During the bulk fermentation the dough starts to produce CO_2 and gain in taste and texture.

鬆弛／發酵

麵糰在造型前需要鬆弛麵筋，時間按不同食譜而異。一般情況，麵糰分割後不會立即造型，所以會預先造雛型，然後置於枱上作鬆弛程序。因這時可讓麵糰鬆弛但不會過度伸展而失去韌度。切記麵糰在鬆身時應蓋上保鮮紙，否則它的表面變乾而呈龜裂狀。

造型

秤重量和分體

量重和麵糰分體可用手或機器處理。用手把麵糰分體，最重要是不能拉扯和撕裂麵糰。同時，如果用手分開，烘焙師應嘗試把每個麵糰均一，盡可能接近所訂定的目標重量，以及避免數小塊疊加一起。

造型方法

這裏有無限方法替麵包造型。每國家的省份均有其之獨特處。最基本的造型是橄欖（卵形）和圓形。法國棒子便是把卵形麵糰伸展長度而成的橄欖形。每個造型可造出不同尺吋，烘焙師常常練習麵包造型，才可以創出個人風格。儘管造型有很多基本手法和原則供參考，個別烘焙師經練習一段時間後，就會訂出自己習性的麵包造型了。

RESTING

The resting time depends on recipes and allows the dough to rest before the shaping. In general, the dough should not be shaped right after being divided; it will often be pre-shaped for proper resting on the table. This time of rest allows the dough to relax and not being over stretched. Remember to always cover your dough while resting or else the dough envelop will dry and form a crust.

SHAPING

WEIGHING AND DIVIDING

Weighing and dividing can be done by hand or by machine. While dividing the dough by hand it is important not to pull on the dough and not to tear the dough. Also, if it is divided by hand, the baker should try to weigh homogeneous pieces of dough, as close as possible of the targeted weight and avoid adding several small pieces together.

SHAPING METHODS

There are unlimited ways to shape bread. Each region in each country has its own specialty and shape. The basic shapes are the oval loaf and the round loaf. The baguette is a derived of the oval loaf, stretched in length. Each shape can then be done in any sizes. Shaping is most likely the exercise that the baker needs to practice the most in order to find his own style. Although there are some basic movements and principles in shaping, individuals tend to develop their own habits after a while.

造型訣竅

　　麵包造型時，有許多因素需要考慮，工作枱是正確造型的主要元素。有些食物安全認證組織曾告誡不正確使用工作枱，多項具爭議性的研究指出含天然細菌的木枱，比沒有天然細菌的膠枱造型有正面影響，反而膠枱缺乏自然反抗力。所以，我建議使用木枱原因。它可卸除因烘焙師搓揉麵糰所產生的壓力。長遠而言，如烘焙師在雲石或不銹鋼枱長時間工作，吸取了許多反震力，令到手臂和肩胛受損。至於工作枱的高度更是烘焙師的背部受到傷害的關鍵。

KEY POINTS DURING SHAPING

There are several factors to take into account while shaping bread. The table you will work on is one of the key elements in the ability to shape bread correctly. Once declared not proper to use by some HACCP bodies, there was controversial studies showing that natural bacteria in wood were having a positive impact rather than none present in plastic boards, being without natural defenses. Therefore, I am recommending to use a wooden table for different reasons, one of which being the pressure created by the baker on the table that is partially absorbed by the wood. On a long term prospect, if the baker shapes on a marble or stainless steel table, the shock absorption by the body will be harmful to arms and shoulders. The height in which you work is also very important for your back.

究竟要用多少力度壓麵糰才適合？因應麵糰類別和造型而決定。換句話說，這有賴麵糰韌度或發酵方法，才按照麵包造型的鬆緊度而採用適當壓力。壓力是否用得恰當會直接影響到烘焙製品的質感和形態。造型時，必須把麵糰的空氣壓出，否則烘焙後的麵糰會出現大氣孔，不夠細緻。

一般而言，麵糰的接合處必須置於麵包底部，除了那些不用造型或翻轉的包類，因為這點決定麵糰經發酵和烘焙後的穩定性。如麵糰接合處須偏在一邊，處理不當，麵包將會出現爆裂或弄成畸型。如麵糰過度造型，最終出現撕裂或弄斷麵筋。烘焙後賣相欠理想，以及難於上升至最佳水平。

The hand pressure on dough needs to be adapted to the type of dough and shaping. Depending on the dough power and leavening method, the pressure needs to be adapted for a tight shaping or a relaxed shaping. This will influence the texture and shape of the bread once baked. During the shaping, it is important to burst large gas bubbles to prevent having a few very large holes in your baked bread.

As a general rule, the welding point must be perfectly under the loaf, except for loaves that are not shaped or turned over. This point will determine the stability of the shape during proofing and baking. Should the welding point be slightly on the side; the bread will open on the welding point and give deformed products. Over shaping a piece of dough will result in tearing it and eventually breaking its gluten threads. The bread will be less attractive after baking and might have difficulties to rise to optimal level.

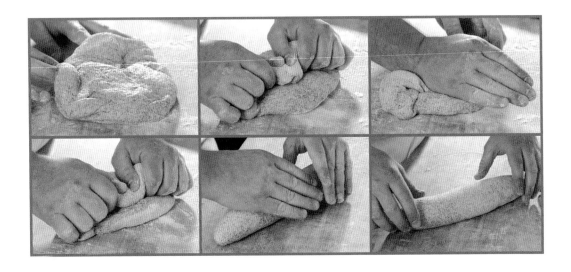

最後發酵

造型後，烘焙師按麵糰情況而決定直接送進發酵爐或緩慢發酵。這裏有控制氣壓的特別房間，以操控相對長時間的發酵過程。在家裏只有冰箱取代發酵爐，雖不理想，但仍可使用，或是烘焙師可把它蓋上保鮮紙，以防止麵糰變硬，然後放進膠箱而置室溫下作長時期發酵。

真正把麵糰入爐前的最後步驟便是最後發酵。烘焙師憑個人感覺和判斷該何時在包面劃紋繼而進行烘焙。在最後發酵時，重點是容許有少許麵包皮外層略乾，以便留下清晰劃紋。如果麵糰的外層過濕，劃紋時，刮刀會黏貼在麵糰不夠利落。

書中的最後發酵時間只作參考，製作時間取決於其溫度、濕度和大小和種類而進行。

PROOFING

Once the dough is shaped, the baker might have decided to conduct its dough directly in the proofer or he might have decided to have a slow fermentation. There are special rooms with controlled atmosphere to control the fermentation over a relatively long period of time. However, at home, you can use your fridge as a replacement. Not optimal, but it works. Or else, the baker might have decided to conduct its dough on a long fermentation at room temperature, covered under a plastic sheet to avoid crusting.

The final proofing is the last step before the actual baking. The baker's senses are sole judges on whether or not the time is right to score and bake a loaf. An important point of the final proofing is allowing the crust to dry a little in order to do clean scores. If the outer layer of the bread is too wet, the blade will stick to the dough while cutting.

The proofing times indicated in this book are for reference. It will depend largely on the temperature and humidity level as well as the size and kind of bread you produce.

刻紋

把麵包刻紋可用不同類型的刀片如鬚刨刀片、特製的麵包刀片或其他利刀劃紋。烘焙師下刀的角度、劃紋深度和數量，需按照發酵度和麵糰類別而為。一般而言，烘焙師會先觀察和觸摸麵糰後才作決定。劃紋時，烘焙師站好，然後充滿信心地俐落下刀。在劃紋前，確保每個麵糰微微分離，不會黏貼在發酵布。當麵糰放在布上，容許有足夠空間讓麵包伸展直至顏色一致。

SCORING

There are different types of blades such as razor blades, special bread blades or other very sharp knifes used to score bread. The angle of the blade, the deepness of scoring and the amount of scoring are depending on the proofing level and on the type of dough the baker has done. Mostly, it is the baker that judges these factors while looking and touching the loaves before scoring. When scoring, the gesture of the baker must be well thought and confident; a score must be done in one stroke. Before scoring, ensure to lift slightly each loaf to make sure that they are not stuck on the fermenting cloth. When placing the loaves on the cloth, allow enough space for the bread to expand and to have a uniform coloration.

烘焙

焗爐類型和溫度

烘焙成功的第一步，預先調校正確的焗爐溫度。市面有許多不同類型的焗爐，如要獲得優質麵包，建議選用有石床的焗爐。因為它能直接與麵糰接觸，熱力直透產品，能助其產生香味、優質鬆脆的外皮、合適體積和吸引的賣相。熱量平衡是指焗爐的熱力均勻。許多時，熱力會自然消失，因接近爐門的溫度較低；所以烘焙師必會把麵包移動，務求各麵包的糊精同質着色，達致色澤均一。書中所指的爐溫和烘焙時間會因應焗爐型號和入爐的麵包數量而調節。

排氣功能

許多在家使用的商業用焗爐是沒有排氣功能。這功能讓蒸氣保存於爐內或可借活門釋放熱氣。

在家烘焙時，最慣常的做法是在入爐前於麵糰噴水。而在爐內直接噴水並不是最好的方法，因會縮減焗爐的壽命。另一烘焙方法，預先燒熱粗陶器壺，並確保有足夠深度容納麵糰，開始烘焙，讓麵糰置於壺內，直至最後階段，把麵包放在外，出現所需顏色。狂熱的家庭烘焙師會把金屬器皿翻轉於石碟上，蓋着麵包，再噴上蒸氣。

BAKING

TYPES OF OVEN AND TEMPERATURES

The first step of a successful baking is to prepare the oven at the right temperature ahead of time. Many different types of oven are available on the market. To obtain a better quality loaf, an oven with a stone bed is recommended. The direct contact of the dough on the stone gives direct heat transfer to the bread and the product will develop all of its aromas, a great crust, a good volume and an attractive shape. The heat balance is the distribution of the heat within the oven. Most of the time, by natural heat loss, the temperature near the doors of the oven is lower, therefore the baker must move the loaves around during baking to obtain a homogeneous coloration of the dextrin. The temperatures and the baking times indicated in this book may vary; it will depend largely on the type of oven you use and on the quantity of bread you produce.

THE EXHAUST FUNCTION

Typically, most commercial ovens you have at home are not equipped with an exhaust function. The functionality of the exhaust is simply to keep steam inside the oven or to release it through a valve.

For home baking, it is a common practice to spray your loaf with water before baking. Spraying directly in the oven might not be the best for your oven's duration. Another way to bake is to preheat a stoneware pot deep enough to accommodate your loaf, start the baking of your loaf in it to later finish proper coloration outside the pot. Extreme home bakers using stone plates will flip a metal container over their loaf and spray steam in it.

烘焙石

在家使用烘焙石，不論你烘焙比薩（披薩）或麵包，能令產品有很大的不同。該石可以跟抗熱功能的精製磚（建造營火焗爐的那種）同類便可。此外，較慣常用的石是菫青石。花崗石和板岩都不錯，當然無論你用任何石，都要沒有上釉那些。

厚度約半吋是最適合。石越厚，預熱時間也越長。如用氣體焗爐，需要用45分鐘加熱，直至熱力均勻。要是用電焗爐，就要60分鐘了。如焗爐的底火和面火，烘焙石能直接置於爐底，底部熱力過強，你需要把石頭放在架上並置於爐的最底部，避免熱力過於集中。

每一塊石，無論人造或天然的，都擁有某程度的濕度，你需要讓它慢慢變乾，否則首次用後會呈現裂紋，甚至斷裂。所以烘焙你的石前，你需要漸進式把它加熱。例如，1小時達至50℃，接着的1小時至100℃，隨後1小時後至150℃，最後半小時達200℃。無需購買有品牌兼昂貴的烘焙石，只要到本地石店，要求他們按麵包中的焗爐尺碼裁出所需石塊便可。

把麵包直接放入已燒熱的石塊，需要用小木板或爐架。選一塊與烘焙石尺碼相若的，灑少許米粉或麵粉避免黏底，然後自信地把已發酵的麵包滑入石上，待烘焙完成後，可用那塊木板或爐架取出麵包。

BAKING STONE

Using a stone in home baking makes a big difference to your end product, whether you bake a pizza or a loaf of bread. The stone can be the same nature as heat resistant "engineered" brick (used to build wood fire oven). Another common natural stone used for baking is cordierite. Granite and slate are also fine; of course any stone used to bake should not be glazed.

A thickness of about half an inch will be appropriate. The thicker the stone, the more time it will take to heat up. If you use a gas oven, it should take a total of about 45 minutes to bring the stone to even heat. And if you are using an electrical oven, it will take about 60 minutes. In an oven with bottom and top heat control, the stone can be directly on the bottom of the oven, but if the heat is too strong from the bottom, you might need to keep your stone on a grid at the lower level in your oven to avoid a too intense heat.

Every stone, either man made or natural have a certain degree of moisture in them and you will need to "dry" them or else the stone will crack or break at first usage. So before baking with your stone, you have to warm it up gradually to "temper" it. For example 1 hour at 50°C followed by 1 hour at 100°C, followed by one hour at 150°C degrees and finally half hour at 200°C. There is no need to get an expensive stone with brand name, visit a local stone shop and ask them to cut a slab of your desired stone for the size of your oven.

To place your bread directly on the heated stone you will need a small wooden board or baker's peel. Choose a size that fits your stone. Simply dust your wood with fine semolina or flour to avoid sticking and place your proofed loaf on it and slide it confidently on the stone. You can use the same wooden board or peel to bring out your baked breads.

烘焙過程

　　當麵包開始烘焙，爐的中心溫度由92℃至96℃之間。酵母會在50℃停止活動，及後，麵包會開始在焗爐內膨脹，這亦解釋了為何入爐後要注入蒸氣，因可阻止外層變乾，好讓麵包在60℃時，澱粉的膠凝作用使泡沫（麵糰）變成海綿狀（麵包）。澱粉酵粒子把澱粉轉化為糊，再變作醣份，麵包香脆堅硬的外層便在這時形成，因外層的水份開始被蒸發掉。溫度達115℃時，幫助麵包帶出香味。有趣的是，麵包中心的溫度反而永遠過不了100℃（水的沸點）。

THE BAKING PROCESS

　　When the loaf is starting to bake, the temperature of the core increases between 92°C to 96°C. The yeast stops its activity at around 50°C, until then, the bread is having its 'oven spring' and continues to further expand–that is why it is important to inject steam when placing the bread in the oven; it will allow the crust to expand, rather than drying out limiting its expansion. The gelatinization of the starch cells, transforming the foam (dough) into a sponge (bread), is happening at 60°C. Alpha-amylases then converts the starch into dextrin and then into sugars. The crust is then created as water is starting to evaporate in the outer layer of the bread. The crust gives its flavor and body to the loaf with the Maillard reaction starting at 115°C. It is interesting to note that the inner temperature of the bread (the crumb) will virtually never pass 100°C– the boiling point of water.

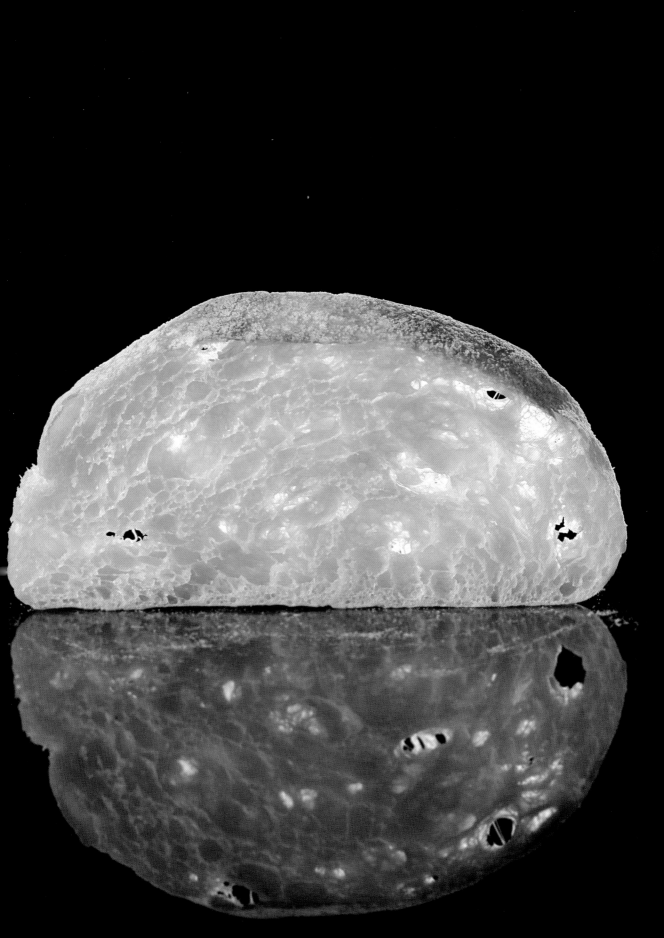

享用　　ENJOYING

當麵包從焗爐拉出來是最令人歡欣喜悅的時間，麵包需要在麵包籃或架子上呼吸。濕度會立即從麵包滲出，約佔其重量約1.5%。所以，如果麵包放在密封的籃子內，它會變得潮濕。除此之外，應該避免極度差異的溫度，不要把熱騰騰的麵包放入冰箱。

控制麵包的品質，我們可用酸鹼度計算其酸性，在酸鹼度表，酸度是以1代表，鹼則是14，7屬於中位數。麵種的酸鹼度是3.5至4.5。正常沒有加入麵種的麵包大概是5.7。溫度會影響我們計算酸鹼值，需一些工具才能電子化地量度。

When the heavenly time of pulling your baked loaves out of the oven, the bread will need to breath in aerated basket or shelves. The level of humidity coming out of the bread straight after baking represents about 1.5% of its own weight. Therefore, if the bread is in closed basket, it will become soggy right away. Additionally, extreme difference of temperature should be avoided, i.e.: don't cool your hot bread in the fridge.

To control the quality of baked bread, we can measure its acidity by measuring the pH. On the pH scale, acidic will be 1 and alkaline will be 14; 7 being neutral. The value of sourdough dough is of 3.5 to 4.5 and regular bread dough without sourdough should be about 5.7. The temperature affects pH measurement, but temperature is electronically leveled using certain tools.

麵糰和麵包的冷藏

從物流目的或生產計劃的目的而言，冷藏生麵糰或半生麵糰是可行的，但麵包師的觀點，麵包需要每天生產比較新鮮一點。

一般人認為-25℃的溫度能保存麵糰不會損害酵母細胞；然而一些調查報告顯示，溫度降至-30℃影響到麵糰的穩定性。除了貯存溫度，冷凍麵糰的重點是其冷凍速度和熔化速度的相互關係，這兩者和冰粒結晶的成因有密切關係。

冰晶是影響冷凍麵糰的主要因素。冰粒的形成視乎晶核及用以使麵糰回暖至鬆弛狀態的熱力穩定

FREEZING DOUGH AND BREAD

For logistical purpose or production planning, freezing raw dough or par baked bread is a possibility; however from an artisan point of view; bread should be produced fresh every day.

It is generally believed that -25°C is adequate to preserve the dough from any physical damages to the yeast cells. However, studies have shown that temperatures below -30°C reveal the real stability of dough. Beside the storage temperature, the major issues of frozen dough is the speed of cooling and the speed of thawing since both of them are related to the formation of ice crystal.

Ice crystals are the main factor affecting quality in dough freezing. The formation of ice will depend on the competition between the ice nucleation and the thawing that needs to allow the dough to relax back to a thermodynamically stable state. The main concern is

性。在冷凍過程，水扮演很重要的角色，水份子轉化為冰粒胚體，例如麵包是完全烘焗和冷凍，其外層的水份含量便少於內層，所以這裏會有少量的水份從外層釋出。

另一方面，濕度較高的環境即麵包內層，經長時期冷藏過程下，水晶粒會形成和膨脹，迫出硬的外層。待融化，內層的冰晶溶化，便會回到原來的形態，麵包外層將會碎掉。利用急劇性的冷藏設施能確保有較好的冷藏過程。

the role of water during the freezing process. Naturally, a water molecule migrates toward the ice embryos. For example if a loaf of bread is fully baked and frozen, the crust will contain less water than the crumb, therefore there will be a minimal migration of water from the crust.

On the other hand, in the higher humidity environment of the crumb, under a long freezing process, the water crystals will form and expand, pushing the crust outward. Once thawing, the ice crystals of the crumb will melt and the crumb will be back to its original shape and the crust will be broken and fall into pieces from the crumb. Using a blast freezing facility guarantees a better freezing process.

閔言樂的麵包廚房
Artisan Bread at Grégoire's kitchen

「沒有感覺就不可能做出好麵包！這就是烘焙的藝術！」簡括而言，烘焙麵包是要一連串複雜的反應中取得平衡，包括良好的物理與化學知識，對材料、烘烤、手藝的精細管理，這些因素結合就是我們所謂的烘焙藝術。

"Without feelings, it's impossible to bake good bread. That is the art of baking" To put it simply, baking bread is a complex balancing act between a good knowledge of physics, chemistry, ingredients, baking, artisanal skills and careful management – these factors combine to form an equilibrium that we call the art of baking.

圖例解讀
YOUR BAKING GUIDE

 室溫（°C）
Room Temperature（°C）

 水溫（°C）
Water Temperature（°C）

 麵糰溫度（°C）
Dough Temperature（°C）

 搓揉時間（螺旋形混合器）
Kneading Time (Spiral Mixer)

 麵糰溫度（°C）
Dough Temperature（°C）

 麵糰發酵（指定小時於指定°C）
Bulk Fermentation (hrs. at °C)

 重量（克）
Weighing （gm.）

 醒發時間（小時 / 分鐘）
Rest Time （hrs./mins.）

 造型（用手搓揉成指定形狀）
Shaping （Loaf shape by hand）

 最後發酵（小時）
Final Fermentation （hrs.）

 最初烘焗溫度（°C）
Initial Baking Temperature （°C）

 烘焗溫度（°C）
Baking Temperature （°C）

 烘焗時間（分鐘）
Baking Time （mins.）

A TOUCH OF
SWEETNESS
甜包

藏紅花香草麵包
SAFFRON VANILLA CUCHAULLE

建議 SUGGESTIONS

由於麵糰含有較大量雞蛋、砂糖和脂肪,接近最後烘焙的階段,必須小心掌控爐溫,因其很快會上色。同時,烘焙前,確保爐內乾爽,沒有預先放入蒸氣。你可以在接近完成烘焗時,把爐門半開。

With a fairly high amount of eggs, sugar and fat, it is important to watch the oven temperatures near the end of baking as the loaves will gain quick coloring. Also ensure to bake the Cuchaulle in a dry oven, without prior steam. You might want to keep your oven's door ajar near the end of baking.

63℃	A: 10 mins(S) B: 5 mins(F)	26℃	30 mins	350 gm

15 mins	spiral	45 mins	180℃	180℃	35 mins

材料 INGREDIENTS		比例 RATIO
480 克 55 號麵粉	480 gm Flour type 55	100%
100 克 雞蛋	100 gm Egg	20.83%
100 毫升 清水	100 ml Water	20.83%
60 克 牛油	60 gm Butter	12 .5%
50 克 白砂糖	50 gm White sugar	10.42%
8 克 乾酵母	8 gm Dry yeast	1.67%
10 克 鹽	10 gm Salt	2.08%
15 條藏紅花雌蕊	15 pcs Saffron pistil	
1 支雲呢拿豆	1 pc Vanilla bean	
糖霜適量（灑面）	Icing sugar for dusting	
白砂糖適量（灑面）	White sugar for dusting	
清水適量	Water	
麵粉適量（灑面）	Dusting flour	

製法 METHOD

1 / 雲呢拿豆剖開，挖出種籽。
2 / 混合所有材料，攪拌成麵糰，麵糰拌滑後加鹽，繼續攪拌3分鐘。
3 / 麵糰鬆弛後，造型如棍狀，再捲成螺旋紋。
4 / 放入已鋪有矽膠蓆的焗盤。
5 / 最後發酵完成後，在麵糰上掃點清水，並灑滿白砂糖。
6 / 灑上大量糖霜，要使用沒有添加粟粉的糖霜。
7 / 焗爐內，不用放蒸氣，並需打開排氣活門，烘焗至金黃色便可。

1. Cut the vanilla bean lengthwise and scrap the vanilla seeds in the dough.
2. Knead the dough with all the ingredients and add the salt 3 minutes prior kneading ends.
3. After the resting time, shape the dough like baguette and roll it as a snail.
4. Place on baking tray with silicon paper.
5. Once fully proofed, brush the loaves with water and sprinkle completely with white sugar.
6. Dust heavily with icing sugar. Make sure to use regular icing sugar without starch added.
7. Bake without steam and open exhaust until golden brown.

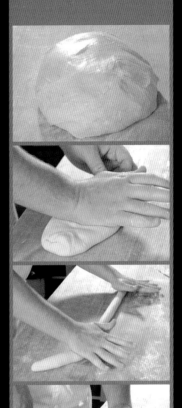

甜包 A touch of sweetness

49

乾果長麵包
THICK DRIED FRUIT LOAF

63℃ | A: 15mins(S) B: 2mins(F) | 27℃ | 1 hr | 5x120 gm

15 mins | round | 2 hrs | 200℃ | 195℃ | 45 mins

材料	INGREDIENTS	比例RATIO
450 克 65 號麵粉	450 gm Flour type 65	100%
120 克 黑麥麵粉	120 gm Rye flour	26.67%
35 克 芝麻	35 gm Sesame seeds	6.14%
40 克 小麥胚芽	40 gm Wheat germs	7.02%
12 克 鹽	12 gm Salt	2.11%
40 克 奶粉	40 gm Milk powder	7.02%
12 克 乾酵母	12 gm Dry yeast	2.11%
400 毫升 清水	400 ml Water	70.18%
麵粉適量（灑面）	Dusting flour	

混合乾果	Fruit mix
30 克 乾杏脯	30 gm Dried apricots
30 克 乾西梅	30 gm Dried prunes
30 克 乾無花果	30 gm Dried figs
30 克 原粒榛子	30 gm Whole hazelnuts
30 克 開邊核桃仁	30 gm Walnut halves
30 克 原粒杏仁	30 gm Whole almonds

製法 METHOD

1/ 混合乾果切成大粒，果仁則保持原粒。
2/ 除了鹽和混合乾果外，混合所有材料，攪拌成麵糰。
3/ 在攪拌過程的最後3分鐘，加入鹽續攪拌完成，然後慢速混入乾果。
4/ 初步發酵完成後，秤重量和讓其鬆弛。
5/ 小心做成圓形，置放在已灑麵粉的長木質發酵籃。
6/ 一旦發酵完成，翻麵糰置於已灑粉的木板上，並在每個麵糰刻紋，再滑落於烘焙石上。
7/ 放入蒸氣後烘焙，當麵糰開始變色，打開排氣活門至完成。

1. Cut all the dry fruits in large chunks and leave the nuts whole.
2. Knead the dough with all ingredients except the salt and the fruit mix.
3. Add the salt 3 minutes prior the end of mixing time. Once the dough is fully kneaded, add the fruits in slow speed.
4. After the bulk fermentation, weight the dough and allow resting.
5. Carefully shape the rounds and place them in a floured long wooden proofing basket.
6. Once fully proofed, flip the bread on a floured board, score each round once and slide the loaf on the baking stone.
7. Bake with steam; open the exhaust once coloring starts.

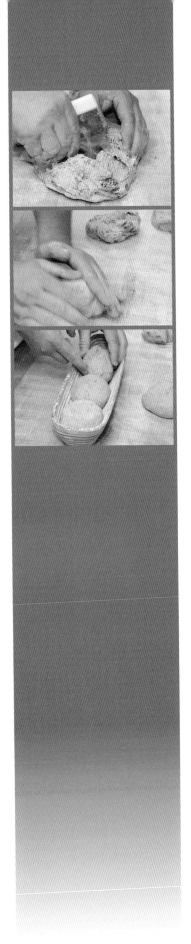

建議 SUGGESTIONS

鑑於麵糰含高份量的乾果和果仁，需要較長的發酵時間。乾果更帶出高含量的天然糖，令麵包很快就變色。

The dough will take a longer fermentation time due to its heavy content of fruits and nuts. The loaf will be gaining color faster due to the high content of natural sugar brought in the dough from the dry fruits.

60℃   A: 12 mins(S) / B: 6 mins(F)   24℃   1.5 hr   450 gm

15 mins   flat   1.5 hrs   200℃   190℃   45 mins

材料	INGREDIENTS	比例 RATIO
250 克 55 號麵粉	250 gm Flour type 55	100%
250 克 770 號黑麥麵粉	250 gm Rye flour type 770	100%
30 克 流質麥芽糖	30 gm Liquid malt	6%
5 克 乾酵母	5 gm Dry yeast	1%
275 克 發酵麵糰	275 gm Fermented dough	55%
450 毫升 清水	450 ml Water	90%
17 克 鹽	17 gm Salt	3.4%
310 克 乾西梅	310 gm Dried prune	62%
160 克 半粒核桃	160 gm Walnut halves	32%

製法 METHOD

1／ 所有材料除核桃和西梅外混合，搓揉成麵糰。
2／ 在攪拌過程完成前3分鐘，加入鹽繼續搓揉。
3／ 麵糰鬆弛後，把麵糰初步造型成扁平厚板狀。
4／ 再造成平長方形，放在已灑粉的布上。
5／ 一旦最後發酵完成後，將麵糰直接滑落在烘焙石上。
6／ 烘焙時放入蒸氣，烘至金黃色；當轉色時立即打開排氣活門。

1. Knead the dough with all the ingredients except the walnuts and the prune.
2. Add the salt 3 minutes prior kneading ends.
3. After the resting time, pre shape the dough in flat slabs.
4. Shape the breads in flat rectangles and place on a floured cloth.
5. Once fully proofed, slide the loaf directly on baking stone.
6. Bake until golden brown with steam; open exhaust once coloring starts.

建議 SUGGESTIONS

由於西梅會帶給麵包的天然深色澤和更多糖份，所以烘焙後，它比一般麵包深色。西梅開半，便能增加西梅塊的份量，把麵包切開時，可讓更多機會見到它們。

已發酵的麵糰是指它早在一日前置於室溫發酵，然後存放於冰箱。如果你沒有預先製作這麵糰，可以製造一小塊，方法非常簡單，只要用上150克 65 號的麵粉，100毫升清水，並與2 至3克 乾酵母，搓揉成糰，置室溫下發酵2小時，放冰箱貯存，直至翌日應用。

Prunes will bring a natural darker color to the bread as well as some extra sugar, giving the loaves an extra darker color after baking. Simply cut the prunes in halves will give you larger chunks of fruit in each slice and as well, it will allow you to keep any potential prune pits out of the bread!

The fermented dough is a piece of dough you have left from the day before and stored in the fridge. If you don't have it on hand, you can produce a small dough. To produce the fermented dough, you can use 150 gm of flour type 65 and 100 ml of water with 2 or 3 grams of yeast, knead it, let it ferment for 2 hours at room temperature and leave it in the fridge until the next day before using.

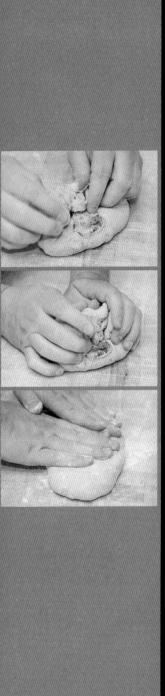

如希望葡萄乾較濕潤，而麵包增添風味，可把葡萄乾用茶或法國阿瑪涅克白蘭地酒或蘋果汁浸軟。這款麵包裏的碧根果仁和葡萄乾，與任何芝士都很配合。

Soak the raisins in tea, Armagnac or apple juice to add extra moisture and flavor to the bread. This bread goes very well with any cheese, complimented by pecan nuts and raisins.

葡萄乾碧根果仁麵包
PECAN AND RASIN BOULE

63℃   A: 10 mins(S)   27℃   1 hr   450 gm
       B: 6 mins(F)

15 mins   round   3 hrs   225℃   200℃   35 mins

材料	INGREDIENTS	比例 RATIO
650 克 65 號麵粉	650 gm Flour type 65	100%
75 克 全麥麵粉	75 gm Whole wheat flour	11.54%
40 克 黑麥	40 gm Rye flour	6.15%
10 克 乾酵母	10 gm Dry yeast	1.31%
30 克 小麥胚芽	30 gm Wheat germs	3.92%
20 克 鹽	20 gm Salt	2.61%
580 毫升 清水	580 ml Water	75.82%
250 克 葡萄乾	250 gm Raisins	32.68%
150 克 碧根果仁	150 gm Pecan nuts	19.61%
麵粉適量（灑面）	Dusting flour	

製法 METHOD

1/ 把所有材料除鹽、葡萄乾和碧根果仁外混合，搓揉成麵糰。
2/ 在攪拌過程完成前 3 分鐘，加入鹽繼續搓揉。
3/ 最後，改以慢速拌入葡萄乾和碧根果仁。
4/ 麵糰初步發酵後，秤重量和讓其鬆弛。
5/ 把麵糰造型，放在已灑麵粉的木製籃繼續發酵。
6/ 當最後發酵完成，把木籃翻轉在已灑粉的板上，再直接滑落於烘焙石上。
7/ 烘焙時放入蒸氣；當麵糰剛開始轉色，立即打開排氣活門，烘至深金黃色。

1. Knead all the ingredients except salt, raisins and pecan nuts.
2. Add the salt 3 minutes prior the end of mixing time.
3. Thoroughly mix the raisins and pecan nuts at the end of the kneading in slow speed.
4. After the bulk fermentation, weigh the dough and allow resting.
5. Shape the Boule and place them in floured proofing wooden basket.
6. Once proofed, turn the baskets on a floured board to slide your loaf directly on the baking stone.
7. Bake with steam and open the exhaust once the bread starts coloring. Bake until dark golden brown.

黑朱古力酸櫻桃麵包
DARK CHOCOLATE & SOUR CHERRY BREAD

63℃   A: 10 mins(S)   27℃   1 hr   450 gm
B: 8 mins(F)

15 mins   round   1.5 hrs   225℃   200℃   40 mins

材料	INGREDIENTS	比例 RATIO
700 克 65 號麵粉	700 gm Flour type 65	100%
125 克 黑麥	125 gm Rye flour	17.86%
70 克 朱古力粉	70 gm Cocoa powder	8.48%
60 克 牛油	60 gm Butter	7.27%
12 克 乾酵母	12 gm Dry yeast	1.45%
28 克 鹽	28 gm Salt	3.39%
620 毫升 清水	620 ml Water	75.15%
190 克 黑朱古力塊	190 gm Dark chocolate chunks	
140 克 冷藏酸櫻桃	140 gm Frozen sour cherries	

製法 METHOD

1 / 除鹽、酸櫻桃和朱古力碎外，把所有材料混合，搓揉成糰。
2 / 在攪拌過程完成前3分鐘，加入鹽繼續搓揉，加入酸櫻桃和朱古力以慢速混合。
3 / 待其鬆弛，預先造型成圓橄欖球狀，續讓它鬆弛。
4 / 造成圓橄欖形，放置在已灑麵粉的發酵布上。
5 / 發酵一旦完成，確認它不黏貼麵包布，在包面上用小刀劃 "X"。
6 / 烘焙時放入蒸氣；待烘焙20分鐘後打開排氣活門至完成。

1. Knead the dough with all the ingredients except the salt, sour cherries and chopped chocolate.
2. Add the salt 3 minutes before the end of kneading; add the sour cherries and chocolate in slow speed, at the end of kneading.
3. Allow resting, pre-shape round loaves and let the dough rest.
4. Shape round loaves and place them on a floured cloth.
5. Once fully proofed, make sure they don't stick to the cloth and give an X cut on the top.
6. Bake with steam; open the exhaust after about 20 minutes of baking.

建議 SUGGESTIONS

早餐時，把這麵包抹上鹹牛油烘暖享用，是一天的完美開始。應使用哈咕油含量最少60% 的黑朱古力粒。

At breakfast, I found this bread served warm and toasted with some salty butter to be the perfect start of the day! Use dark chocolate of at least 60% cocoa content or more to snatch that chocolate bite up!

國王麵包
KING'S BREAD

63℃   A: 8 mins(S) / B: 10mins(F)   25℃   45 mins   450 gm

15 mins   round   1 hr   200℃   185℃   55 mins

建議 SUGGESTIONS

如要呈現清晰的切割口，在進行切割和烘焙時，確保麵糰沒有過度發酵。技巧是於掃蛋液前，讓麵糰外層略乾燥，因過濕的麵糰會黏貼住刀片。

To obtain a clean cut, ensure your loaf is not over proofed when cutting and baking. Once egg washed, allow the crust to dry a little before cutting in order to have clean cuts and avoiding wet dough sticking to the blade.

材料	INGREDIENTS	比例 RATIO
500 克 45 號麵粉	500 gm Flour type 45	100%
300 毫升 牛奶	300 ml Milk	60%
12 克 乾酵母	12 gm Dry yeast	2.4%
40 克 白砂糖	40 gm White sugar	8%
10 克 鹽	10 gm Salt	2%
30 克 雞蛋	30 gm Egg	6%
1 個 檸檬皮	1 pc Lemon zest	
90 克 牛油	90 gm Butter	18%
35 克 杏仁醬	35 gm Almond paste	7%
1 支雲呢拿豆	1 pc Vanilla bean	
蛋液適量	Egg wash	
朱古力粒適量	Cocoa nibs	

製法 METHOD

1/ 混合所有材料，除牛油和鹽外，搓揉成麵糰。
2/ 在攪拌過程完成前3分鐘，加入鹽和牛油繼續搓揉。
3/ 蓋上保鮮紙，讓麵糰發酵。
4/ 秤重量和初步造型成圓形狀，讓麵糰鬆弛。
5/ 造成圓卷形和置放在已墊烘焙紙的焗盤上。
6/ 進行最後發酵過程。
7/ 用手指在麵糰的中央處戳一個孔至75% 深度。
8/ 掃上蛋液，並從外至內沿周邊下刀切割，灑上朱古力粒。
9/ 烘焙時放入蒸氣，焗至金黃色；待開始轉顏色，打開排氣活門。

1. Knead the dough with all ingredients except butter and salt.
2. Add the salt and butter 3 minutes prior the end of mixing time.
3. Let the dough rest as it is, covered with a plastic film for the bulk fermentation.
4. Weigh the dough and pre-shape in round shape; allow resting.
5. Shape in round loaves and place them on a baking tray lined with baking paper.
6. Proof to the final fermentation.
7. Gently punch a whole in the center of the loaf with a finger to 75% of the depth.
8. Egg wash and cut the loaf from the outside to the inside, all around. Sprinkle cocoa nibs.
9. Bake with steam until golden brown; open the exhaust once coloring starts.

椰糖牛油粒圓包
Coconut Palm Sugar Roll

63℃   A: 12 mins(S)   26℃   45mins   50 gm
B: 5 mins(F)

10 mins   buns   1.5 hrs   200℃   190℃   25 mins

建議 SUGGESTIONS

麵糰含的椰子成份相當高，在麵糰看見椰油是正常的。用冷牛油做甘寶粒，質感完美。此食譜用了 Masarang 印尼椰糖，它能帶出獨特香味，加強與甘寶形成強烈對比。再者，它的餘韻與椰子的清香配合得天衣無縫。

Due to the high coconut content in the dough, you might see traces of coconut oil on the dough, it's normal. Use cold butter to make your crumble and you'll obtain the perfect texture. I used Masarang palm sugar from Indonesia; it gives a unique flavor and intense contrast to the crumble. Moreover, it undertone perfectly the mellow aromas of coconut.

材料	INGREDIENTS	比例 RATIO
420 克 45 號麵粉	420 gm Flour type 45	100%
8 克 乾酵母	8 gm Dry yeast	1.9%
40 克 椰子粉	40 gm Coconut milk powder	9.52%
25 克 椰糠	25 gm Dessicated coconut	5.95%
25 克 白砂糖	25 gm White sugar	5.95%
12 克 鹽	12 gm Salt	2.86%
60 克 牛油	60 gm Butter	14.29%
280 毫升 椰汁	280 ml Coconut milk	66.67%
蛋液適量（掃面）	Egg wash	
麵粉適量（灑面）	Dusting flour	

甘寶	Crumble
50 克 牛油	50 Butter
70 克 椰糖	70 Palm sugar
65 克 自發粉	65 Self raising flour
25 克 椰糠	25 Dessicated coconut

製法 METHOD

1/ 甘寶材料混合一起，壓入隔篩，做成粒粒，置冰箱備用。
2/ 搓揉麵糰，在攪拌過程完成前3分鐘，才加入鹽，繼續搓揉。
3/ 麵糰完成初步發酵後，秤重量，讓其鬆弛。
4/ 造型，放在已鋪焗紙的焗盤上。
5/ 當完成最後發酵，掃上蛋液，灑上甘寶粒粒於麵糰面。
6/ 不用放蒸氣，打開排氣活門直接烘焙至金黃色。
7/ 出爐後待涼，灑上糖霜。

1. Prepare the crumble by mixing all the ingredients together and pass it through a grid to make crumbles. Reserve in the fridge.
2. Knead the dough and add the salt 3 minutes prior kneading ends.
3. After the bulk fermentation, weigh the dough and allow resting.
4. Shape the bun and place them on a baking tray lined with baking paper.
5. Once fully proofed, apply egg wash and sprinkle crumbles on top of the buns.
6. Bake until golden brown without steam and with open exhaust.
7. Dust icing sugar once cooled.

蜜餞薑粒皇家麵包
Candied Ginger Pan D'oro

63℃　　A: 14 mins(S)　25℃　　one night　50 gm
　　　　B: 6 mins(F)

20 mins　　round　　3 hrs　　180℃　　180℃　　25 mins

建議 SUGGESTIONS

這食譜只需很少量的酵母，經長時間發酵和稀流麵糰，便可製造出非常濕潤和可口的麵包。蜜餞薑粒皇家麵包是傳統年末節慶而準備的麵包。

This recipe calls for very little yeast, a very long fermentation and a very liquid dough making it very moist and delicious. Pan d'Oro is traditionally made for the end of year festivities.

材料	INGREDIENTS	比例 RATIO
280 克 45 號麵粉	280 gm Flour type 45	100%
80 毫升 牛奶	80 ml Milk	28.57%
90 克 白砂糖	90 gm White sugar	32.14%
110 克 牛油	110 gm Butter	39.29%
180 克 雞蛋	180 gm Egg	64.29%
8 克 乾酵母	8 gm Dry yeast	2.86%
5 克 鹽	5 gm Salt	1.79%
1/2 個檸檬皮	half pc Lemon zest	
2 湯匙杏仁酒	2 tbsp Amaretto	
1 湯匙洋槐蜜糖	1 tbsp Accacia honey	
75 克 蜜餞薑粒	75 gm Diced candied ginger	
白砂糖適量	White sugar	
糖霜適量	Icing sugar	

製法 METHOD

1 / 搓揉麵糰，在接近完成攪拌的3分鐘前加入鹽拌勻。
2 / 在室溫下鬆弛半小時，摺疊一次。
3 / 把麵糰放在膠器皿，蓋上保鮮紙，放入5℃冰箱過一夜。
4 / 翌日，再次摺疊麵糰一次，秤重量，初步做出圓形。
5 / 讓其鬆弛和做成小卷麵包，放在已鋪烘焙紙的金屬圈。
6 / 置室溫下發酵3小時。
7 / 掃上清水，灑上砂糖和大量糖霜，放入已開啟排氣活門的焗爐烘焙。

1. Knead the dough and add the salt 3 minutes prior kneading ends.
2. Let it rest at room temperature for half an hour and fold it once.
3. Keep the dough in a covered plastic container overnight in the fridge at 5℃.
4. The next day, fold the dough again, weigh it and preshape it in round shape.
5. Allow resting and shape the rolls. Place them in metal ring lined with paper.
6. Proof at room temperature for about 3 hours.
7. Brush water, dust sugar and icing sugar heavily and bake with exhaust open.

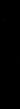

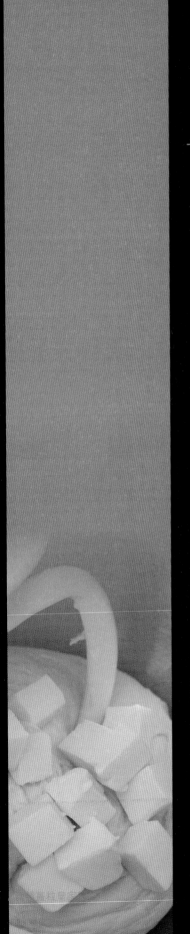

材料	INGREDIENTS	比例 RATIO
牛油撻	Brioche	
500 克 45 號麵粉	500 gm Flour type 45	100%
12 克 白砂糖	12 gm White sugar	2.4%
130 克 牛油	130 gm Butter	26%
180 克 蛋黃	180 gm Egg yolk	36%
12 克 鹽	12 gm Salt	2.4%
80 毫升 鮮牛奶	80 ml Fresh milk	16%
10 克 乾酵母	10 gm Dry yeast	2%
12 克 麥芽糖	12 gm Malt	2.4%
蛋液適量（掃面）	Egg wash	
麵粉適量（灑面）	Dusting flour	
啤梨杏仁餡	Frangipane	
150 克 牛油	150 gm Butter	
150 克 白砂糖	150 gm White sugar	
150 克 雞蛋	150 gm Eggs	
150 克 杏仁粉	150 gm Almond powder	
40 克 自發粉	40 gm Self raising flour	
190 克 糕餅忌廉	190 gm Pastry cream	
鹽 1 少撮	A pinch of salt	
水煮梨適量	Poached pears	

建議 SUGGESTIONS

把啤梨浸於有肉桂、丁香和八角的啤梨汁，添加啤梨撻的風味。再掃上少許啤梨和雲呢拿果醬，能為水果表面增加額外光澤。

Poach your pears in pear juice with some cinnamon, cloves and star anise to add extra flavors to your tart. Brush a little pear and vanilla jam on the fruits to give it an extra shine.

63℃　　　　　A: 12 mins(S)　　25℃　　　1 hr　　　90 gm
　　　　　　　B: 5 mins(F)

10 mins　　flat round　　30 mins　　200℃　　190℃　　25 mins

製法 METHOD

1/　搓揉麵糰，在攪拌過程完成前3分鐘，加入鹽續攪拌完成。
2/　蓋上保鮮紙，讓麵糰發酵。
3/　秤重量，預先做出圓形狀，讓麵糰鬆弛。
4/　在已灑麵粉的枱上，把牛油撻麵糰碾擀至約1厘米厚，鈒出一個直徑約20厘米的圓形。
5/　放在已鋪焗餅紙的焗餅盤上，讓其最後發酵至3/4倍。
6/　在麵糰中央處輕輕壓下，除邊緣的2厘米外，掃上蛋液。
7/　在撻的中央，抹上啤梨杏仁餡。
8/　加入切成片的水煮梨於啤梨杏仁餡。
9/　不用放入蒸氣於焗爐，但需開啟排氣活門，烘焗至金黃色。
10/　烘焙完成，立即徐徐掃上啤梨或杏脯果醬，再灑上糖霜。

1. Knead the dough and add the salt 3 minutes prior kneading ends.
2. Let the dough rest as it is, covered with a plastic film for the bulk fermentation.
3. Weigh the dough and pre-shape in round shape; allow resting.
4. Roll the tart on a floured table at around 1cm thick and cut a disc of dough of about 20 cm.
5. Place them on a baking tray lined with baking paper. Allow to proof to 3/4.
6. Press gently the middle part of your dough, leaving a 2cm rim to be egg washed.
7. Spread some frangipane mixture on the center of the tart.
8. Add your sliced pears on the frangipane mixture.
9. Bake without steam and with open exhaust until golden brown.
10. Once baked, brush the pear lightly with pear or apricot marmalade, sprinkle with icing sugar.

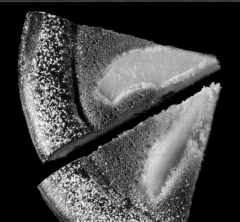

榛子朱古脆卷
HAZELNUT PRALINE FLAKY ROLLS

 61℃ A: 8 mins(S) B: 8 mins(F) 24℃ 30 mins 3x80 gm

 15 mins snail rolls 1 hrs 200℃ 190℃ 24 mins

材料	INGREDIENTS	比例 RATIO
500 克 45 號麵粉	500 gm Flour type 45	100%
250 毫升 鮮牛奶	250 ml Fresh milk	50%
15 克 乾酵母	15 gm Dry yeast	3%
50 克 白砂糖	50 gm White sugar	10%
10 克 鹽	10 gm Salt	2%
25 克 雞蛋	25 gm Egg	5%
1 個檸檬皮	1 pc Lemon zest	0.2%
200 克 牛油（夾心）	200 gm Butter for folding	40%
100 克 榛子（烘焙）	100 gm Hazelnut, roasted	
麵粉適量（灑面）	Dusting flour	
蛋液適量（掃面）	Egg wash	

餡料	Filling
90 克 牛油	90 gm Butter
50 克 榛子醬	50 gm Hazelnut praline paste
120 克 黃砂糖	120 gm Brown sugar

製法 METHOD

1/ 預備餡料，把牛油和榛子醬混合至幼滑，加入黃砂糖，置室溫下備用。
2/ 搓揉麵糰，搓揉完成前3分鐘，加入鹽繼續搓。
3/ 待發酵後，秤重量和碾擀至成正方形，置冰箱內約3小時。
4/ 把牛油夾心碾擀為1厘米厚的片裝，放在麵糰中央的位置。將周邊麵糰向內摺覆，包裹牛油。
5/ 再把麵糰徐徐碾成1.5厘米厚，簡單摺疊。
6/ 重複（步驟5）2次，用保鮮紙包好麵糰，使用前放進冰箱待2小時。
7/ 在工作枱上灑粉，碾擀麵糰約3毫米厚，抹上榛子醬和灑上榛子碎。
8/ 在麵糰的角位掃上一次蛋液，捲成螺旋紋。把麵糰切割，將3個小麵糰貼合成一排，放在已鋪紙的焗盤上。
9/ 最後發酵完成後，掃點蛋液，灑上榛子碎，放入焗爐放蒸氣和關閉排氣活門，烘至金黃色。

1. Prepare the filling; mix the butter and hazelnut praline into a creamy texture, add the brown sugar. Keep aside, at room temperature.
2. Knead the dough and add the salt 3 minutes prior kneading ends.
3. After the bulk fermentation, weigh the dough and flatten it into a square. Allow cooling in the fridge for about 3 hours.
4. Roll the folding butter into a sheet of about 1cm in thickness and place it in the middle of your dough. Fold the side towards the center to encase the butter in the dough.
5. Roll the dough gently to 1.5 cm thickness and give it a simple fold.
6. Repeat the step 5 another two times and wrap your dough in plastic film. Keep it in the fridge for 2 hours before using.
7. Roll the dough on a floured surface, to 3 mm in thickness. Spread the hazelnut filling and sprinkle with the chopped hazelnuts.
8. Apply egg wash once on one edge of the dough and roll into snail. Cut and place them in row of 3 pieces stuck together, on a baking tray lined with baking paper.
9. Once fully proofed, apply egg wash, sprinkle chopped hazelnut and bake until golden brown with steam and closed exhaust.

建議 SUGGESTIONS

榛子脆卷的質素與所用的榛子醬相關，要買純淨的榛子醬，因沒有添加糖，能保持原有的味道。入爐前，在已掃蛋液的榛子麵包卷上，加點鏡面醬、糖霜或只是簡單地多加一些已烘焙的榛子便可。

These flaky hazelnut rolls are as good as the hazelnut praline paste quality you'll use in the making. Try to find pure hazelnut paste, without sugar added to obtain a maximum of flavor. You can finish these rolls with glazing, icing sugar or simply add more roasted hazelnuts on the egg wash, before baking.

珍珠糖及朱古力忌廉麵包
PEARL SUGAR & CHOCOLATE BREAD

A: 10 mins(S)
B: 6 mins(F)

 26℃

 1 hr

 350 gm

 1.5 hrs

 190℃

 190℃

 35 mins

INGREDIENTS	比例 RATIO
600 gm Flour type 55	**100%**
10 gm Dry yeast	**1.67%**
30 gm Milk powder (full fat)	**5%**
80 gm White sugar	**13.33%**
10 gm Salt	**1.67%**
60 gm Egg	**10%**
250 ml Fresh milk	**41.67%**
150 ml Liquid cream	**25%**
130 gm Chocolate 70%	
60 gm Icing sugar	
50 gm Pearl sugar	
Dusting flour	
Egg wash	

製法 METHOD

1/ 把朱古力磨碎，與糖霜混合，作釀餡備用。
2/ 搓揉麵糰，完成前3分鐘，加入鹽。
3/ 讓麵糰鬆弛，然後擀碾成長方形1.5厘厚米和30厘米潤的。
4/ 掃上蛋液，把朱古力混合物均勻地掃在麵糰上。
5/ 將麵糰捲成如蝸牛似的螺旋紋，按模具的大小切成塊。
6/ 把長卷切半，再把兩半麵糰扭在一起，待發酵。
7/ 當發酵完成，掃上蛋液和灑上珍珠糖。
8/ 不用放蒸氣，直至入焗爐，打開排氣活門，焗至金黃色。

1. Ground your chocolate into a powder and mix with icing sugar; reserve for the stuffing.
2. Knead the dough and add the salt 3 minutes prior kneading ends.
3. Allow resting; roll the dough in rectangular shape of about 1.5 cm thick and 30 cm in width .
4. Egg wash and spread the cocoa mixture evenly over the dough.
5. Roll the dough as a snail, cut it into chunks to fit your mould in length.
6. Cut your roll in halves and twist the two halves together. Proof.
7. Once proofed, egg wash and sprinkle the pearl sugar.
8. Bake without steam and open exhaust until golden brown.

建議 SUGGESTIONS

把優質純黑朱古力磨成粉狀作餡料。如要製作出漂亮的旋渦效果，應放足夠份量的朱古力粉混合物於麵包裏。用珍珠糖能保持外層白皙和酥脆，如使用一般的白糖就會熔掉和變焦糖。

Simply take a nice dark chocolate and ground it into powder for the filling. Use a good quantity of chocolate powder mixture inside the bread to create a nice swirl effect. Pearl sugar will remain white and crisp whereas regular sugar would melt and caramelise.

甘薯琥珀核桃包

SWEET POTATO & CARAMELISED CHINESE WALNUT "COCOTTE"

| 63℃ | A: 12 mins(S) B: 6 mins(F) | 25℃ | 45 mins | 50 gm |

| 10 mins | round rolls | 2.5 hrs | 190℃ | 180℃ | 35 mins |

材料	INGREDIENTS	比例 RATIO
420 克 45 號麵粉	420 gm Flour type 45	100%
30 克 蛋黃	30gm Egg yolk	7.14%
50 克 雞蛋	50gm Egg	11.90%
20 毫升 牛奶	20 ml Milk	4.76%
5 克 乾酵母	5gm Dry yeast	1.19%
8 克 鹽	8gm Salt	1.90%
75 克 黃砂糖	75gm Brown sugar	17.86%
115 克 牛油	115gm Butter	27.38%
240 克 熟番薯茸	240gm Cooked/mashed sweet potato	57.14%
1 個檸檬皮	1 pc Lemon zest	
60 克 琥珀核桃碎	60gm Chinese caramelized walnuts, chopped	
麵粉適量（灑面）	Dusting flour	
蛋液適量	Egg wash	

建議 SUGGESTIONS

用焗爐烘烤番薯可避免它們有過多水份，待涼後，在加入麵糰前先壓成幼滑薯茸。
按照番薯的狀況，如感覺麵糰有點乾，可加入少許清水調節。

Bake your yellow sweet potatoes in the oven will avoid making them watery. Once cold, mash them smoothly before adding to the dough. Depending on the potatoes, you might need to add a little water if the dough is too hard.

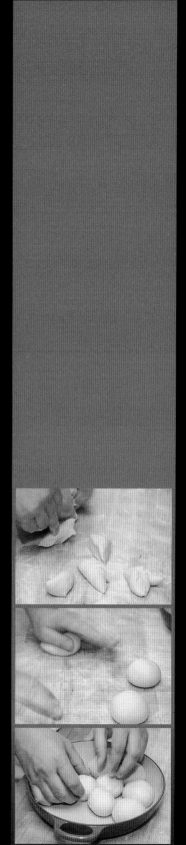

製法 METHOD

1/ 搓揉麵糰，在攪拌過程完成前3分鐘加入鹽，續攪拌至完成。琥珀核桃作裝飾，備用。
2/ 麵糰鬆弛，秤重量。初步造成卷狀和再次鬆弛。
3/ 造成圓形，放在已掃油的煎鍋。
4/ 將小麵卷發酵，掃上蛋液和灑上琥珀核桃。
5/ 放入蒸氣，烘焙至金黃色。當麵包正開始轉色，打開排氣活門。
6/ 可選擇放上金箔作裝飾。

1. Knead the dough; add the salt 3 minutes prior kneading ends. Keep walnuts for garnish.
2. Allow resting and weigh the dough; pre-shape it in rolls and rest the dough again.
3. Shape the round rolls and place them in your greased cocotte pan.
4. Proof the buns; apply egg wash and sprinkle with caramelised walnuts.
5. Bake with steam until golden brown; open exhaust once coloration starts.
6. Decorate with optional gold leaves.

THE ARROGANCE OF
SALT
鹹包

德國番茄辣椒蝴蝶包
TOMATO CHILI PRETZEL

建議 SUGGESTIONS

為保麵包有嚼頭而且外型挺拔，麵糰需要比其他麵糰硬，並在短時間內造型。

工作時確保不要徒手觸及鹼水，必須穿上手套和保護衣物。鹼水是腐蝕性物質，會傷害皮膚。如果接觸到，立即用肥皂於清水下沖洗。經烘焙後，鹼水的侵蝕力便失去，可以安全進食。

The dough needs to be harder than other dough and conducted in a faster way in order to obtain chewy and good looking pretzel.

Be sure not to touch the lye with your bare hand and wear gloves and proper protective cloth while working with it. Lye is corrosive and will cause skin injury. If in contact, wash under water with soap right away. Once baked, lye is no longer corrosive and is safe to be consumed.

63℃	A: 10 mins(S) B: 6 mins(F)	27℃	1 hr	50 gm	

5 mins	pretzel	15 mins	220℃	200℃	25 mins

材料	INGREDIENTS	比例 RATIO
600 克 65 號麵粉	600 gm Flour type 65	100%
15 克 乾酵母	15 gm Dry yeast	2.5%
250 毫升 清水	250 ml Water	41.67%
10 克 鹽	10 gm Salt	1.67%
15 毫升 橄欖油	15 ml Olive oil	2.5%
5 克 白砂糖	5 gm White sugar	0.83%
60 克 濃縮番茄	60 gm Tomato concentrate	10%
乾刁草碎適量	SQ Dried chilli flakes	
鹼水適量	SQ Lye (Caustic soda)	
麵粉適量（灑面）	Dusting flour	

製法 METHOD

1/ 搓揉麵糰，在攪拌過程的最後3分鐘，加鹽續攪拌至完成。
2/ 鬆弛完成後，把麵糰造成蝴蝶包的交叉狀，讓其發酵。
3/ 準備一盤鹼水，穿上保護衣、手套和眼鏡。
4/ 把蝴蝶包浸入鹼水，翻轉，用有柄隔篩撈出，除去多餘的鹼水。
5/ 放入已鋪烘焙紙的焗盤，灑上乾辣椒。
6/ 立即放入焗爐烘焗至深啡色，不需放蒸氣於爐內，但要開啟排氣活門。

1. Knead the dough and add the salt 3 minutes prior kneading ends.
2. After resting time, shape the bread into pretzel shape and allow them to proof.
3. Prepare your lye bath in a container. Wear protective cloth, gloves and glasses.
4. Deep your pretzel in the lye, turn it over and take it out with a strainer ladle to remove the excess lye.
5. Place on a tray lined with baking paper and sprinkle with dried chili.
6. Bake immediately until dark brown, without steam and with exhaust opened.

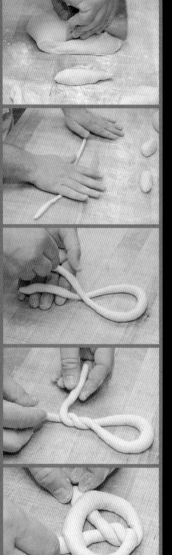

辣椒格魯耶芝士條
GRUYÈRE CHEESE & PAPRIKA TWIST

建議 SUGGESTIONS

你可以用不同芝士做麵包條,例如藍芝士或意大利巴馬芝士,但格魯耶芝士的味道會貼合。有橄欖油的麵糰會帶給麵包條額外酥脆感。如果它們不夠鬆脆,可把它留在已降溫的焗爐長一點時間才取出,讓它乾身一點,有較佳效果。如欲獲得較堅硬和粗糙的外觀,烘焙前不用讓這些芝士條發酵。

You can use different cheese to make the twist, such as blue cheese or parmesan, but Gruyere fits the job very well. The olive oil in the dough brings a little extra crunch to the sticks. If they're not crunchy, let them dry a little longer in a cooler oven. We don't let these sticks proofing before baking as we want to obtain that tough and rough look.

63℃　　　A: 12 mins(S)　　26℃　　　1 hr　　60 gm
　　　　　　B: 6 mins(F)

15 mins　　twisted sticks　　5 mins　　200℃　　180℃　　35 mins

材料	INGREDIENTS	比例 RATIO
410 克 65 號麵粉	410 gm Flour type 65	100%
6 克 乾酵母	6 gm Dry yeast	1.46%
310 毫升 清水	310 ml Water	75.61%
14 克 鹽	14 gm Salt	3.41%
12 毫升 橄欖油	12 ml Olive oil	2.93%
格魯耶芝士適量	SQ Gruyere cheese	
紅椒粉適量	SQ Paprika powder	
麵粉適量（灑面）	Dusting flour	
蛋液適量（掃面）	Egg wash	

製法 METHOD

1/ 搓揉麵糰，在攪拌過程的最後3分鐘，加入鹽繼續攪拌，完成後拌入芝士和紅椒粉。將麵糰置冰箱鬆弛約1小時。
2/ 鬆弛完成，麵糰擀成3毫米至5毫米厚。
3/ 在麵糰上均勻地掃蛋液，灑上大量的碎格魯耶芝士。
4/ 灑上紅椒粉，翻轉厚片麵糰，在另一面再灑上紅椒粉。
5/ 用滾輪式切麵刀把麵片切成3厘米濶的長條。
6/ 對摺麵糰條，扭在一起，放入已鋪焗餅紙的焗盤上，進行最後發酵。
7/ 一旦麵糰鬆弛完成，在焗爐內放入蒸氣，烘焙至金黃色，待麵包轉色時，打開排氣活門。

1. Knead the dough and add the salt 3 minutes prior kneading ends. Keep the cheese and paprika for the finishing. Keep the dough in the fridge for the rest time (1 hour).
2. After resting time, roll the dough between 3 mm to 5 mm thick.
3. Brush egg wash all over the dough and sprinkle heavily with grated gruyere cheese.
4. Sprinkle paprika powder and flip the slab of dough. Repeat the operation on the other side.
5. Using a cutting wheel, cut stripes of 3 cm in width.
6. Fold the stripes by half lengthwise and twist them; place on a baking tray with baking paper and proof.
7. Once rested, bake until golden brown with steam; open exhaust once coloring starts.

黑胡椒茴香脆條
BLACK PEPPER & FENNEL TARALLI

材料	INGREDIENTS	比例RATIO
500 克 00 號麵粉	500 gm Flour type 00	100%
15 克 鹽	15 gm Salt	3%
150 毫升 白酒	150 ml White wine	30%
120 毫升 橄欖油	120 ml Olive oil	24%
黑胡椒適量（壓碎）	SQ Black pepper, cracked	
茴香籽適量（乾燥）	SQ Fennel seeds, dried	
麵粉適量（灑面）	Dusting flour	
清水適量（煮麵包用）	Water for boiling	

建議 SUGGESTIONS

這傳統南意大利「麵包條」，做法與賓果包相同，方法是在烘焙前把麵糰用水煮片刻。這種包通常給造成環形，但我覺得其他形狀更有趣。如果你覺得質感過於緊密，可加少許酵母或發粉。這些脆包可與任何香料配搭，也可做成甜的。適合配葡萄酒或湯來吃。

This classic "bread stick" from the south of Italy follows the same way of bagel's making, being boiled in water for a short while before baking. The taralli are usually shaped into closed rings, but I find other shapes attractive. If you find the texture too compact, you can add a little yeast or baking powder. The taralli can be flavored with any spices and can also be made in sweet version. These crispy sticks are great enjoyed with a glass of wine or with any soups.

55℃   A: 8 mins(S)   23℃   30 mins   30 gm
B: 4mins(F)

no rest time   'U' shape sticks   none   200℃   200℃   35 mins

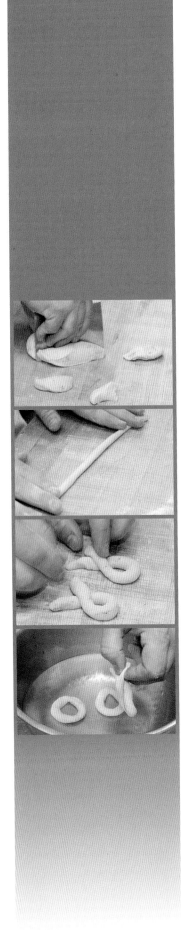

製法 METHOD

1/ 把所有材料混合，攪拌成麵糰，放下讓其鬆弛。
2/ 麵糰切成小塊，碾平至條狀，每條長約20厘米。
3/ 將清水和1大湯匙鹽置於鍋中煮熱。
4/ 把條狀麵糰放入熱水中煮幾分鐘，直至浮起。
5/ 用隔篩撈出麵糰，放在烘焙焗盤上造型。
6/ 不放蒸氣，關閉排氣活門，烘焙，當麵包開始轉色，立即開啟排氣活門。
7/ 想做到非常酥脆和乾身，讓麵包條停留在已轉低溫的爐中長一點時間才取出。

1. Knead the dough with all ingredients and allow resting.
2. Cut the pieces and roll as bread stick of approximately 20 cm in length.
3. Boil water in a pot with a large tablespoon of salt.
4. Deep the pieces of dough in the hot water for a few minutes, until they float.
5. Using a strainer, take them out and place on a baking tray, forming different shapes.
6. Bake without steam and closed exhaust. Open the exhaust once coloring starts.
7. For very crispy and dry taralli, let them dry at a lower temperature for a longer time.

法國紅辣椒薄脆
PIMENT D'ESPELETTE LAVOSH

材料	INGREDIENTS	比例 RATIO
520 克 65 號麵粉	520 gm Flour type 65	100%
260 克 45 號麵粉	260 gm Flour type 45	50%
10 克 白砂糖	10 gm White sugar	1.28%
20 克 鹽	20 gm Salt	2.56%
12 克 乾酵母	12 gm Dry yeast	1.54%
150 毫升 鮮牛奶	150 ml Fresh milk	19.23%
190 毫升 清水	190 ml Water	24.36%
110 克 牛油（夾心）	110 gm Butter for folding	14.1%
法式紅椒粉適量	SQ Espelette pepper	
麵粉適量（灑面）	Dusting flour	

建議 SUGGESTIONS

脆片做法是越薄越好。按你的口味，可加點海鹽或香辛料；但要小心，因為你開始享用這美味的薄脆便停不了。薄脆不會在發酵時膨脹，但麵包會在烘焙過程產生少量氣泡。長時間進行鬆弛過程，能防止麵糰縮回原狀，特別是在切割成方形或三角形後，盡量保持不變形。

The thinner the better! You can also add some sea salt or other spices, up to your taste buds. But beware, once you start eating these Lavosh, you won't be able to stop! The Lavosh won't rise during the fermentation time, but the bread will eventually generate a few bubbles during the baking process. The long resting time is simply to prevent the dough from retracting itself too much, especially when cutting shapes such as squares or rectangles, to keep their shape as sharp as possible.

| 61℃ | A: 8 mins(S)
B: 3mins(F) | 23℃ | 30 mins | - |

| none | flat crisp | 1 hr | 170℃ | 170℃ | 35 mins |

製法 METHOD

1/ 搓揉麵糰，其麵筋不要太堅韌。
2/ 發酵後，碾平麵糰厚約2厘米，蓋上保鮮紙，放入冰箱貯藏2小時。
3/ 夾心牛油放在2張烘焙紙間，碾成薄片。
4/ 牛油放在麵糰的中央，覆蓋麵糰，如牛角包做法。
5/ 擀成2厘米厚，視覺上把麵糰分為3份，將它摺疊成3層。
6/ 擀成2厘米厚，重複摺及擀薄2次。
7/ 包裹麵糰，放入冰箱貯藏1小時。
8/ 把麵糰擀到紙一樣薄，排放在已鋪焗餅紙的焗盤上。
9/ 鬆弛約半小時，徐徐掃上清水，灑上紅椒粉。
10/ 此時可以切割麵糰，或原片留待烘焙後才把它弄碎。
11/ 無需放入蒸氣，關閉排氣活門，然後烘焙，待脆片開始轉色，立即開啟排氣活室。

1. Knead the dough with a not so strong gluten network.
2. After the bulk fermentation, flatten your dough to 2 cm thick and store in the chiller for about 2 hours, covered with a plastic film.
3. Roll your folding butter between 2 pieces of baking paper to obtain a thin butter sheet.
4. Place the butter in the middle of the dough and fold the dough to cover the butter, like in making croissant dough.
5. Roll at 2 cm and visually divide the dough in 3 parts and fold into 3 tiers.
6. Roll again the dough at 2 cm and repeat the folding 2 more times.
7. Wrap the dough and keep it in the fridge for 1 hour.
8. Roll the dough paper-thin and lay it on a baking tray with baking paper.
9. Allow resting for half an hour and lightly brush with water. Sprinkle Espelette pepper.
10. You may cut it now, or leave it as a whole sheet to be broken into rough pieces once baked.
11. Bake without steam and closed exhaust. Open the exhaust once coloring starts.

64℃	A: 10 mins(S) B: 6 mins(F)	26℃	1 hr	25 gm

30 mins	sharp needle	45 mins	200℃	185℃	20 mins

材料	INGREDIENTS	比例 RATIO
麵種	**Starter**	
150 克 55 號麵粉	150 gm Flour type 55	100%
40 克 牛油	40 gm Butter	26.67%
40 克 白砂糖	40 gm White sugar	26.67%
40 克 蛋黃	40 gm Egg yolk	26.67%
45 克 全蛋	45 gm Whole egg	30%
12 克 乾酵母	12 gm Dry yeast	8%
100 毫升 清水	100 ml Water	66.67%
麵糰	**Dough**	
150 克 55 號麵粉	150 gm Flour type 55	100%
15 克 奶粉	15 gm Milk powder	10%
7 克 鹽	7 gm Salt	4.67%
20 克 新鮮蒜頭（切碎）	20 gm Freshly chopped garlic	13.33%
5 克 新鮮番荽	5 gm Freshly chopped parsley	3.33%
蛋液適量（掃面）	Egg wash	

建議 SUGGESTIONS

麵糰半冷藏後方便造型。確保蒜茸和番荽碎是現剁現用，並且不要把它們混合太久。蒜頭的蒜氨酸和番荽的葉綠素都會影響到麵筋的筋度，會令麵包的質感變差。

Semi freezing the dough allows you to obtain very sharp shape. Ensure to use freshly chopped garlic and parsley, as well, don't mix these 2 ingredients for too long. Both, the garlic alliinase and the parsley chlorophyll have been scientifically identified as substances affecting the gluten strength and thus, the crumb of your breads.

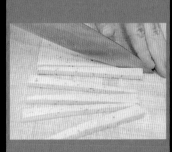

製法 METHOD

1/ 把所有麵種材料混合搓揉成糰，讓其置室溫下發酵1小時。
2/ 加入麵粉和奶粉搓揉麵糰，在攪拌過程的最後3分鐘，加入鹽續攪拌至完成。
3/ 以慢速拌入蒜頭和番荽。
4/ 發酵麵糰後，碾擀麵糰成長方形，約1厘米厚×15厘米闊。
5/ 放已灑麵粉的焗盤，蓋上保鮮紙，貯放冰格30分鐘。
6/ 把麵糰放在已灑麵粉的板上，切成長三角針形。
7/ 放入焗盤，掃上蛋液，進行發酵。
8/ 當發酵完成，再掃一次蛋液，放入蒸氣，然後烘焙，待麵包開始轉色，立即開啟排氣活門。

1. Knead the starter and let it ferment for 1 hour at room temperature.
2. Knead the dough and add the salt 3 minutes before kneading ends.
3. Add the garlic and parsley at the end of kneading in slow speed.
4. After the bulk time, roll the dough in 1 cm thick rectangle of about 15cm in width.
5. Lay the dough on a floured tray, cover with a plastic film and store in the freezer for 30 minutes.
6. Place the dough on a floured cutting board and cut the long triangle needle.
7. Put them on a baking tray and egg wash once; proof.
8. Once fully proofed, egg wash once more and bake with steam. Open the

煙肉香草包
BACON & HERBS EPI

64℃　　　　　A: 10 mins(S)　　26℃　　　1.5 hrs　　　350 gm
　　　　　　　B: 8 mins(F)

15 mins　　épi baguette　　1.5 hrs　　235℃　　215℃　　35 mins

材料	INGREDIENTS	比例RATIO
速成麵種	**Poolish**	
200 克 65 號麵粉	**200 gm Flour type 65**	
200 毫升清水	**200 ml Water**	
3 克乾酵母	**3 gm Dry yeast**	
麵糰	**Dough**	
400 克 65 號麵粉	**400 gm Flour type 65**	100%
400 克熟速成麵種	**400 gm Ripe poolish**	100%
2 克乾酵母	**2 gm Dry yeast**	0.5%
15 克鹽	**15 gm Salt**	3.75%
250 毫升清水	**250 ml Water**	62.5%
35 克炒香煙肉	**35 gm Sauteed bacon**	8.75%
4 克雜乾香草	**4 gm Mixed dry herbs**	1%
麵粉適量（灑面）	**Dusting flour**	

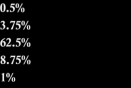

製法 METHOD

1/ 搓揉麵糰，在攪拌過程的最後3分鐘，加入鹽續攪拌完成。
2/ 攪拌過程一旦完成，以慢速加入煙肉和香草。
3/ 鬆弛麵糰，秤重量，把麵糰造型，成長棍狀。
4/ 當鬆弛完成，再次搓揉成長棍狀，放在已灑麵粉的發酵布上發酵。
5/ 用金屬麵糰刮刀把麵糰切幾刀，放在焗盤烘焗或直接置於烘焙石均可。
6/ 放蒸氣於焗爐，關閉排氣活門，然後烘焙，待麵包轉色，立即開啟排氣活門。

1. Knead the dough and add the salt 3 minutes before kneading ends.
2. Add the bacon and herbs in slow speed, once the kneading process is completed.
3. Allow resting and weigh the dough. Pre-shape in baguette.
4. Once rested, shape the dough into baguette and proof on a floured cloth.
5. Cut in epi using a metal dough scraper and bake on tray or directly on stone.
6. Bake with steam and closed exhaust. Open the exhaust once coloring starts.

建議 SUGGESTIONS

在長形棍狀麵糰上灑少許麵粉，繼而碾擀麵糰，可造出粗糙感。把它分扯成小卷，便是宴會上跟朋友歡聚，不可或缺的小吃，非常吸引。

Roll the baguette in a little flour once finished to shape to give it a rustic look. This baguette is convivial and attractive to eat with friends, pulling apart each single roll. To obtain the characteristic shape of the epi, ensure to not over proof your baguette before cutting.

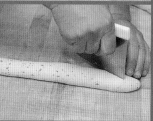

蒜蓉香草俄羅斯軟包
RUSSIAN PAMPUSHKA

材料	INGREDIENTS	比例RATIO
500 克 45 號麵粉	500 gm Flour type 45	100%
140 克 牛油	140 gm Butter	28%
50 克 白砂糖	50 gm White sugar	10%
5 克 鹽	5 gm Salt	1%
10 克 乾酵母	10 gm Dry yeast	2%
275 毫升 鮮牛奶	275 ml Fresh milk	55%
60 克 牛油（熔化）	60 gm Butter, melted	
新鮮蒜頭 1 球（剁碎）	1 head Garlic, fresh, chopped	
新鮮番荽適量（剁碎）	SQ Parsley, fresh, chopped	
麵粉適量（灑面）	Dusting flour	
蛋液適量（掃面）	Egg wash	

64℃ A: 10 mins(S) 25℃ 1 hr 30 gm
      B: 6 mins(F)

15 mins bunch of round 45 mins 215℃ 200℃ 35 mins

製法 METHOD

1/ 牛油放小鍋內煮熔，在起煙前離火，加入鮮蒜頭碎和番荽碎，放在和暖處讓其味道泡出。

2/ 搓揉麵糰，在攪拌過程的最後3分鐘，加鹽續至攪拌完成。

3/ 初步發酵麵糰後，秤重量和進行鬆弛。

4/ 造成圓卷，把圓卷的邊沿互相緊貼，置放在已塗油的焗盤。

5/ 發酵麵糰，待完成後掃上蛋液。

6/ 放入蒸氣於爐內，關閉排氣活門，然後烘焗，待麵包開始轉色，立即開啟排氣活門。

7/ 麵包出爐後，趁熱抹上香蒜牛油待片刻，趁暖享用。

1. Melt the butter in a pot and just before boiling point, add the chopped raw garlic and the parsley. Let it sit in a warm place to infuse the flavors.

2. Knead the dough and add the salt 3 minutes prior kneading ends.

3. After the bulk fermentation, weigh and allow resting.

4. Form the round rolls and place them, touching each others, in an oiled baking tray.

5. Proof the dough and brush egg wash once.

6. Bake with steam and closed exhaust. Open the exhaust once coloring starts.

7. Once baked and while hot, brush the garlic butter a few times on the bread and enjoy while still warm.

建議 SUGGESTIONS

這麵包要趁暖吃，若涼了，包面的牛油會凝結。大量的蒜茸和番荽可增強味道。這款軟包在俄羅斯的傳統烹飪是與俄羅斯湯伴吃。如果你想麵包多點牛油，烘焙前就不要掃蛋液。如此做法，麵包的光澤會減少，但多些牛油會滲於麵包內。

Eating this bread cold will not be as pleasant as warm as the butter topping will be set. Use a lot of garlic and parsley to make the flavors stronger. In Russian classic cooking, the pampushka is usually served with Borsch soup. If you want to have your bread with more butter, don't apply the egg wash before baking. That way, it'll be less shiny, but butter will go deeper in the bread crumb.

新鮮迷迭香方包
FRESH ROSEMARY PAVÉ

64℃

A: 10 mins(S)
B: 8 mins(F)

25℃

1 hr

450 gm

15 mins

rectangle

1.5 hrs

235℃

210℃

40 mins

材料	INGREDIENTS	比例 RATIO
400 克 65 號麵粉	400 gm Flour type 65	100%
6 克 乾酵母	6 gm Dry yeast	1.5%
15 克 鹽	15 gm Salt	3.75%
15 克 新鮮迷迭香	15 gm Fresh rosemary	3.75%
300 毫升 清水	300 ml Water	75%
20 毫升 橄欖油	20 ml Olive oil	5%
優質粗粒小麥粉	Fine semolina	
新鮮迷迭香條適量	Fresh rosemary sprigs	
海鹽適量	Sea salt	

建議 SUGGESTIONS

把新鮮迷迭香略粗切碎，烘焙時把其味道散發，透過發酵過程，麵糰會同時把迷迭香精油完全滲入，如用乾迷迭香便不會達到相同效果。

The roughly chopped fresh rosemary will release its flavor while baked. Through the process of fermentation, the dough will also impregnate itself with the rosemary essential oil. Using dry rosemary wouldn't achieve the same result.

製法 METHOD

1/ 把迷迭香切碎。
2/ 搓揉麵糰，在攪拌過程的最後3分鐘，加入鹽和碎迷迭香，續攪拌完成。
3/ 初步發酵麵糰完成，放在已灑粗粒小麥粉的枱上，分成小粒，秤重量，讓麵糰自然平放成形。
4/ 放在已灑粗粒小麥粉的發酵布上進行最後發酵。
5/ 完成醒發，用手指在麵糰戳孔，灑上優質粗粒小麥粉和海鹽。
6/ 在麵糰插上迷迭香枝葉，放入蒸氣，直接放在烘焙石上烘焗，當麵包轉色立即開啟排氣活門。

1. Chop the rosemary into rough pieces.
2. Knead the dough; add the chopped rosemary and salt 3 minutes prior kneading ends.
3. After the bulk fermentation, place the dough on a table dusted with fine semolina and weight the pieces in natural flatten shape.
4. Place on a cloth dusted with fine semolina to proof.
5. Once fully proofed, make holes with your finger, dust semolina and sea salt.
6. "Plant" rosemary tips on the dough and bake directly on the baking stone, with steam. Open the exhaust once coloring starts.

小洋蔥細葉芹小圓包
BABY ONION & CHERVIL BITES

63℃   A: 10 mins(S) B: 5 mins(F)   25℃   2.5 hrs   15 gm

10 mins   small discs   1.5 hrs   220℃   200℃   18 mins

材料	INGREDIENTS	比例RATIO
400 克 65 號麵粉	400 gm Flour type 65	100%
250 毫升 鮮牛奶	250 ml Fresh milk	62.5%
60 克 牛油	60 gm Butter	15%
20 克 白砂糖	20 gm White Sugar	5%
7 克 乾酵母	7 gm Dry yeast	1.75%
80 克 雞蛋	80 gm Egg	20%
25 克 乾洋蔥片	25 gm Dried onion flake	6.25%
10 克 鹽	10 gm Salt	2.5%
山蘿蔔苗適量	Baby chervil	
蛋液適量（掃面）	Egg wash	

酒浸烤洋蔥	Onion confit
3 個 紅洋蔥	3 pcs Red onions
2 湯匙 牛油	2 tbsp Butter
2 湯匙 橄欖油	2 tbsp Olive oil
150 毫升 梅洛紅酒	150 ml Merlot red wine
2 片 月桂葉	2 pcs Bay leaves
1 茶匙 黃砂糖	1 tsp Brown sugar
2 湯匙 黑醋	2 tbsp Balsamic vinegar
鹽和胡椒粉適量	SQ Salt and pepper

製法 METHOD

1 / 準備酒浸烤洋蔥：先把紅洋蔥去衣和切片。在鍋內，以少許牛油和橄欖油慢煎洋蔥，令其汁液流出。加入紅酒、月桂葉、黃砂糖續慢煮至軟身。最後，加入黑醋、鹽和胡椒，調味。
2 / 搓揉麵糰，在攪拌過程的最後3分鐘，加入鹽續攪拌至完成。
3 / 一旦鬆弛完成，秤重量，把麵糰造出初型，像圓球狀，讓其鬆弛。
4 / 再把麵糰搓圓，完成醒發。
5 / 當麵糰醒發後，掃上蛋液，在焗爐放入蒸氣，然後關閉排氣活門，待烘焗至麵糰轉色，便要開啟排氣活門。
6 / 烘焙完成，把麵包對半切開，塗少許牛油烘烤，在其中央加入足夠份量的酒浸烤洋蔥和山蘿蔔苗便成。

1. Prepare the onion confit; peel and slice the red onions into slices. In a pot, slowly cook the onion with the butter and olive oil and make them sweat. Add the red wine, bay leaves and brown sugar and cook slowly until soft. Finally, add the balsamic vinegar, salt and pepper and adjust the taste.
2. Knead the dough and add the salt 3 minutes before kneading ends.
3. Once rested, weigh the dough and pre-shape into round pieces; allow resting.
4. Shape the round buns and proof completely.
5. Once proofed, brush with egg wash and bake with steam and closed exhaust. Open the exhaust once coloring starts.
6. Once baked, cut them in halves and toast them with a little butter; add a generous amount of confit in the middle and some baby chervil.

建議 SUGGESTIONS

酒浸烤洋蔥的質地應該是像溶化般的而不是呈糊狀。方法是必須以慢火烹煮洋蔥約45分鐘至1小時左右。如果你想增添酸果味道，可加點橙皮豐富其食味。

The onion confit texture should be melting, and not mushy. It is important to cook it on very slow heat for about 45 minutes to 1 hour. To enhance it, you can add a little orange zest if you like citrus flavor.

帕爾馬火腿芝麻菜比薩
Arugula & Parma Ham Pizza

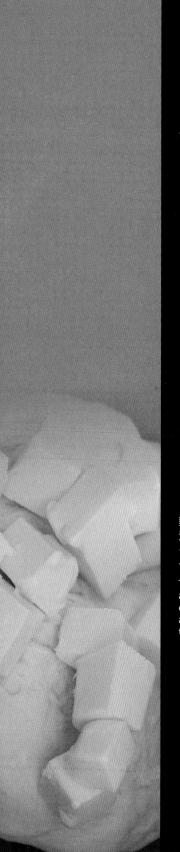

材料	INGREDIENTS	比例 RATIO
250 克 00 號麵粉	250 gm Flour type 00	100%
150 毫升 清水	150 ml Water	60%
4 克 乾酵母	4 gm Dry yeast	1.6%
12 克 海鹽	12 gm Sea salt	4.8%
芝麻菜（火箭菜）適量	SQ Arugula	
帕爾馬火腿片適量	SQ Parma ham slice	
帕爾馬芝士適量	SQ Parmesan cheese	
番茄醬適量	SQ Tomato sauce	
新鮮牛至葉適量	SQ Fresh oregano	
橄欖油適量	SQ Olive oil	
黑胡椒適量	SQ Black pepper	
莫薩里拉芝士適量	SQ Mozzarella cheese	
麵粉適量（灑面）	Dusting flour	

建議 SUGGESTIONS

麵糰的厚薄，悉隨尊便。你可以造薄脆或厚如麵包。想添加豪華感，可在薄餅上滴上少許黑松露油甚至放上黑松露片。00 號的麵粉最適合不過，但其他優質麵粉也可。想令薄餅有脆薄的效果，必須以短時間在非常高溫的焗爐烘烤，這樣可以令薄餅周邊呈深色特點。

The thickness of the dough depends on your liking. You may do it thin and crispy or thicker and more bread-like. To add a little luxurious touch to your pizza, add a drizzle of truffle oil or even freshly slices truffles on top. The flour type oo suits best the purpose, but other flour are also fine to use. A thin pizza should be baked in a very hot oven for a short time, giving these characteristic dark edges to the pizza.

| 64℃ | A: 10 mins(S) B: 8 mins(F) | 26℃ | 1.5 hrs | 200 gm |

| 1 hr | flat round | none | 270℃ | 270℃ | approximately 12 mins |

製法 METHOD

1/ 搓揉麵糰，在攪拌過程的最後3分鐘，加入鹽續攪拌完成。
2/ 初步發酵麵糰完成後，秤重量，分成小糰，搓成球狀。
3/ 蓋上保鮮紙，置室溫下進行鬆弛。
4/ 在鬆弛時間，準備包面的所需材料。
5/ 麵糰放在粗粒小麥粉上，擀成0.5至1厘米厚的圓形。
6/ 確保工作枱上時常有足夠的麵粉，避免麵糰黏貼在枱上。
7/ 麵糰上抹上番茄醬，但醬不要塗到邊上，灑上莫薩里拉芝士碎。
8/ 不用放入蒸氣，開啟排氣活門，直接放在已處高溫的焗爐的烘焙石烤焗。
9/ 當烘焙完成，趁熱把其他剩餘材料，隨意加在薄餅上，完成後放上芝麻菜。

1. Knead the dough and add the salt 3 minutes prior kneading ends.
2. After the bulk fermentation, weigh and form balls of dough.
3. Cover with a plastic film and allow resting at room temperature.
4. During the resting time, prepare all the ingredients needed for the topping.
5. Roll your dough on fine semolina at about 0.5 to 1 cm in thickness.
6. Ensure to always have enough semolina to avoid the dough from sticking to the surface.
7. Spread the tomato sauce, leaving the edges free of sauce. Sprinkle the shredded mozzarella cheese.
8. Bake in a very hot oven directly on your baking stone without steam and open exhaust.
9. Once baked and while hot, add all the other ingredients randomly, finishing the top with the arugula.

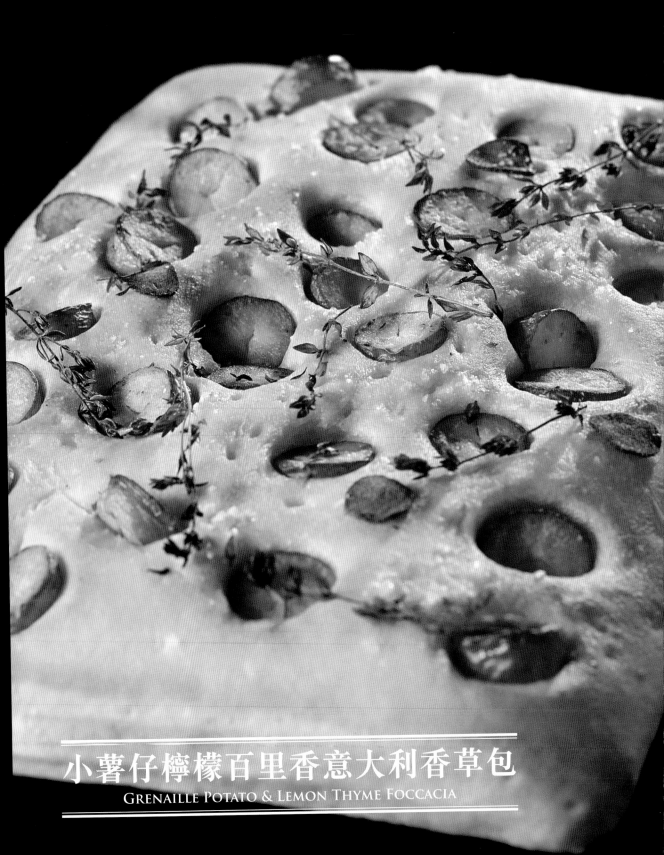

小薯仔檸檬百里香意大利香草包

Grenaille Potato & Lemon Thyme Foccacia

64℃	A: 8 mins(S) B: 6 mins(F)	24℃	1 hr	tray size
none	in tray	45 mins	225℃	200℃	40 mins

建議 SUGGESTIONS

烘焗薯仔而無需用清水，目的是阻止過多水份加入麵糰內。於麵糰加入薯仔可令麵包的澱粉質增加，造出一層香脆的外層，更可令麵包的水份恰到其份。此麵包的最佳食法，便是加入檸檬百里香、橄欖油和海鹽於煙三文魚或配合其他豐盛的意大利菜。

The potatoes are baked without water with the purpose of not adding extra liquid to the dough. The addition of potatoes in the dough will bring an extra starch in the bread which will give a beautiful crust and perfect moisture to the bread. Lemon thyme, olive oil and sea salt are great flavors to go with smoked salmon and all those hearty Italian dishes.

材料	INGREDIENTS	比例RATIO
375 克 45 號麵粉	375 gm Flour type 45	100%
150 毫升 鮮牛奶	150 ml Fresh milk	40%
150 毫升 清水	150 ml Water	40%
40 毫升 橄欖油	40 ml Olive oil	10.67%
12 克 鹽	12 gm Salt	3.2%
15 克 乾酵母	15 gm Dry yeast	4%
375 克 焗薯仔	375 gm Baked potatoes	100%
焗小薯仔適量	SQ Baked grenaille potatoes	
新鮮檸檬百里香適量	SQ Fresh lemon thyme	
海鹽適量	SQ Sea salt	
橄欖油適量	SQ Olive oil	
麵粉適量（灑面）	Dusting flour	

製法 METHOD

1/ 把薯仔連皮放在焗盤，蓋上錫紙，以爐溫200℃焗1小時，直至完全熟透。涼凍後，去掉薯皮，壓成粗薯茸狀。
2/ 烹煮連皮小薯仔，然後涼凍。切成1厘米厚片。
3/ 搓揉麵糰，在攪拌過程的最後3分鐘，加入鹽和粗薯蓉，續至攪拌完成。
4/ 初步發酵麵糰完成，碾平麵糰，放進已掃油的焗盤，進行發酵。
5/ 麵糰完全醒發後，在麵糰隨意以手指戳上多個小孔，然後掃上橄欖油。
6/ 放小薯仔片於麵糰上，輕輕按壓，灑上海鹽和檸檬百里香。
7/ 放蒸氣於爐內，關閉排氣活門，然後烘焙，待麵糰轉色，立即開啟排氣活門。

1. Bake the unpeeled baking potatoes in a tray covered by aluminum foil at 200°C for about 1 hour, until fully cooked. Once cold, peel them and crush them in chunky puree.
2. Cook the grenaille potatoes with their skin and let cool. Cut into 1cm thick slices.
3. Knead the dough and add the salt and the crushed potatoes 3 minutes prior kneading ends.
4. After the bulk fermentation, roll the dough in an oiled tray and proof.
5. Once fully proofed, puncture holes with your fingers randomly and brush olive oil.
6. Place the grenaille potatoes slices on the dough, pressing them a little, and sprinkle sea salt and lemon thyme.
7. Bake with steam and closed exhaust. Open the exhaust once coloring starts.

建議 SUGGESTIONS

Fougasse 香草扁麵包是南部法國的 Foccacia，屬香草包類。你可以在麵糰加入其他材料烘焙或選用不同面料也可。另一個 Fougasse 的版本是在包內釀入不同餡料，同樣美味。

Fougasse is thin flat bread from southern France that resembles Foccacia bread. You can add other ingredients to the dough or use different toppings before baking. Another version of the Fougasse is with stuffing inside the dough.

普羅旺斯法式香草扁麵包
PROVENCAL FOUGASSE

64°C A: 10 mins(S) / B: 6 mins(F) 26°C 1.5 hrs 450 gm

15 mins fougasse 45 mins 235°C 225°C 40 mins

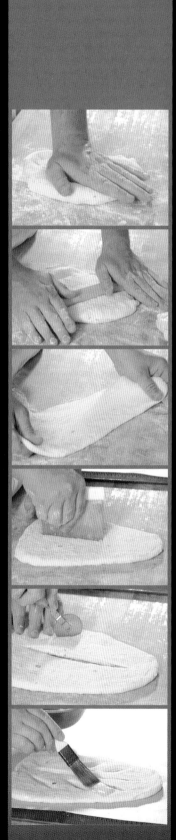

材料 INGREDIENTS		比例RATIO
麵種 Starter		
150 克 65 號麵粉	150 gm Flour type 65	
180 毫升 清水	180 ml Water	
5 克 乾酵母	5 gm Dry yeast	
5 克 白砂糖	5 gm White sugar	
麵糰 Dough		
550 克 65 號麵粉	550 gm Flour type 65	100%
350 毫升 清水	350 ml Water	63.64%
330 克 麵種	330 gm Starter	60%
7 克 乾酵母	7 gm Dry yeast	1.27%
20 克 鹽	20 gm Salt	3.64%
20 毫升 牛奶	20 ml Fresh milk	3.64%
20 毫升 橄欖油	20 ml Olive oil	3.64%
6 克 普羅旺斯乾香草	6 gm Provence herbs, dried	1.09%
50 克 日曬番茄乾	50 gm Chopped sun dry tomato	9.09%
30 克 黑橄欖	30 gm Black olives	5.45%
30 克 Bayonne 火腿	30 gm Bayonne ham	5.45%

製法 METHOD

1 / 麵種材料混合，待發酵置室溫待4~5小時。
2 / 搓揉麵糰，在攪拌過程的最後3分鐘，加入鹽續攪拌完成。
3 / 以慢速加入番茄、橄欖、香草和Bayonne火腿片。
4 / 鬆弛完成後，把麵糰擀成厚長方塊。
5 / 在中間處沿麵糰長度切割一刀，每邊分別再切3刀，揭開切口，放入焗盤。
6 / 放入蒸氣烘焗至金黃色，待麵包開始轉色，立即開啟排氣活門。

1. Mix the starter and let it ferment for 4 to 5 hours at room temperature.
2. Knead the dough and add the salt 3 minutes before kneading ends.
3. Add the tomatoes, olives, herbs and Bayonne ham pieces in slow speed.
4. After the resting times, roll the dough in large rectangle slabs.
5. Give one long cut in the middle and 3 cuts on each side of it. Open the cuts and place on a baking tray.
6. Bake with steam until golden brown. Open the exhaust once coloring starts.

番茄乾意式煙肉包
SUN DRIED TOMATOES & PANCETTA ROLLS

建議 SUGGESTIONS

可在自製番茄醬汁內加入濃縮番茄醬，令其味道更強烈。這醬汁需要用最少份量的水，待烘焙後，可減低麵包之間的裂口或缺口。

Add some concentrated tomato paste in your tomato sauce to make it even more intense. The sauce needs to have the least water content possible in order to diminish the potential gap between layers of bread once baked.

63℃   A: 10 mins(S)   26℃   45 mins   80 gm
B: 6 mins(F)

15 mins   rolls   45 mins   200℃   190℃   35 mins

材料	INGREDIENTS	比例 RATIO
380 克 65 號麵粉	380 gm Flour type 65	100%
150 毫升 牛奶	150 ml Milk	39.47%
150 毫升 清水	150 ml Water	39.47%
40 毫升 橄欖油	40 ml Olive oil	10.53%
15 克 鹽	15 gm Salt	3.95%
10 克 乾酵母	10 gm Dry yeast	2.63%
80 克 日曬番茄乾	80 gm Sun dried tomato	21.05%
番茄醬適量	SQ Tomato sauce	
意大利煙肉適量	SQ Pancetta	
新鮮羅勒適量	SQ Fresh basil	
麵粉適量（灑面）	Dusting flour	
蛋液適量（掃面）	Egg wash	
乾雜香草適量	Dried herbs	

製法 METHOD

1/ 加入鹽和黑胡椒，煮番茄醬，待水分收乾至剩下三分之一左右，即成為濃郁的番茄醬。
2/ 搓揉麵糰，在攪拌過程的最後3分鐘，加入鹽續攪拌至完成，以慢速拌入日曬番茄乾粒。
3/ 鬆弛後，量出麵糰，初步造成圓形。
4/ 待再次醒發後，把麵糰碾成直徑12厘米的圓盤狀，抹上番茄醬，但外周邊1厘米處，不用抹醬。
5/ 加入意大利煙肉和羅勒碎。
6/ 把圓平麵糰捲成卷狀，並在中央部份切割，微微將內部呈現出來，以突出內部層次。
7/ 放入已鋪焗餅紙的焗盤上。
8/ 發酵完成，掃上蛋液和放上意式火煙肉、芝士和羅勒裝飾。
9/ 放入蒸氣，烘焗至金黃色，待麵包轉色便開啟排氣活門。

1. Cook your tomato sauce with salt and black pepper and reduce it of a third to obtain a thicker tomato sauce.
2. Knead the dough and add the salt 3 minutes prior kneading ends. Add the chopped sun dried tomatoes at the end of kneading in slow speed.
3. After resting time, scale the dough and pre-shape in round.
4. Once rested, roll the dough into discs of about 12cm and spread the tomato sauce, leaving a rim of 1cm around the disc of dough.
5. Add the pancetta in pieces and the roughly chopped basil leaves.
6. Fold the disc into a roll and cut the middle part all the way through. Push the inner part to come out slightly in order to expose the inner layers.
7. Place on a baking tray with baking paper and proof.
8. Once fully proofed, brush with egg wash and garnish with pancetta, cheese and basil.
9. Bake until golden brown with steam; open exhaust once coloring starts.

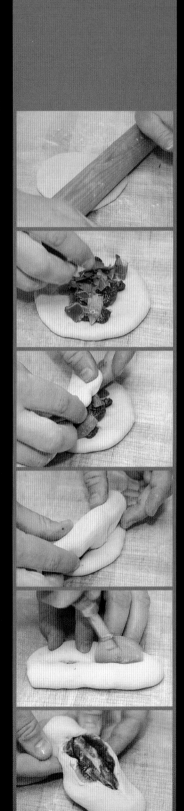

希臘黑橄欖法式麵包
KALAMATA OLIVES BAGUTTE

62℃ · A: 10 mins(S) B: 8 mins(F) · 25℃ · 1hr · 450 gm

20 mins · baguette · 45 mins · 225℃ · 200℃ · 35 mins

材料	INGREDIENTS	比例RATIO
300 克 65 號麵粉	300 gm Flour type 65	100%
50 克 全麥麵粉	50 gm Whole wheat flour	
50 克 黑麥麵粉	50 gm Rye flour	
6 克 乾酵母	6 gm Dry yeast	2%
15 克 小麥胚芽	15 gm Wheat germs	5%
12 克 鹽	12 gm Salt	4%
325 毫升 清水	325 ml Water	81.25%
90 克 Kalamata 橄欖	90 gm Kalamata olives	30%
麵粉適量（灑面）	SQ Dusting flour	

製法 METHOD

1/ 搓揉麵糰，在攪拌過程的最後3分鐘，加鹽續搓至完成。
2/ 一旦攪拌完成，以慢速拌入原粒橄欖。
3/ 當麵糰鬆弛完成後，造型，呈棍子狀。
4/ 麵糰鬆弛後，再次做成棍形，輕輕灑上麵粉。
5/ 放在發酵布上進行發酵麵包。
6/ 發酵完成後，沿麵糰長度，放在烘焙石上。
7/ 放入蒸氣於爐內，烘焙至金黃色，當其開始轉色，開啟排氣活門。

1. Knead the dough; add the salt 3 minutes prior kneading ends.
2. Once the kneading is done, add the whole olives in slow speed.
3. After resting time, pre-shape in baguette.
4. Allow resting and shape the bread into baguettes, slightly dusted with flour.
5. Proof your breads on a floured cloth.
6. Once proofed, score your bread with a long cut and place it on your baking stone.
7. Bake until golden brown with steam; open the exhaust once coloring starts.

建議 SUGGESTIONS

優質 Kalamata 橄欖是帶出特別風味的主要關鍵。選用 Calamata 品牌橄欖和橄欖油，給予你最佳效果和 Kalamata 的真正味道。

The quality of Kalamata olives used is the key to a great flavor in this bread. Using the original Calamata Brand olives, and olive oil for that matter, will help you obtain the finest results and true flavor of Kalamata.

大蘑菇鄉村麵包
PORTOBELLO MUSHROOM COUNTRY BREAD

建議 SUGGESTIONS

此食譜用了意大利大蘑菇，燒法可把其特別香味帶出，與麵包十分配合，尤其用炭燒方法更佳；也可用其他菇類代替。蘑菇包最佳食法是烘烤，然後放上芝士做三文治或法式或意式開邊三文治，配以各款湯類，非常美味。

You can use any type of mushrooms; however, grilled Portobello mushrooms have a specific flavor that mixes well in bread, especially char-grilled. Mushroom bread is great toasted and served warm with any type of soups or melted cheese sandwiches and tartines. Naturally, mushrooms contain a lot of water, therefore the amount of water is reduced in the recipe or else the dough will be too soft.

| 65°C | A: 12 mins(S) B: 8 mins(F) | 26°C | 1 hr | 450 gm |

| 15 mins | round loaf | 45 mins | 235°C | 210°C | 45 mins |

材料	INGREDIENTS	比例 RATIO
650 克 65 號麵粉	650 gm Flour type 65	100%
100 克 全麥麵粉	100 gm Whole wheat flour	15.38%
100 克 黑麥麵粉	100 gm Rye flour	15.38%
12 克 乾酵母	12 gm Dry yeast	1.41%
35 克 小麥胚芽	35 gm Wheat germ	4.12%
20 克 鹽	20 gm Salt	2.35%
550 毫升 清水	550 ml Water	64.71%
300 克 意大利大蘑菇	300 gm Portobello mushroom	35.29%
2 克 黑胡椒碎	2 gm Cracked black pepper	0.24%
麵粉適量（灑面）	Dusting flour	

製法 METHOD

1/ 把新鮮大蘑菇清洗和切厚片，用橄欖油和鹽燒熟。
2/ 搓揉麵糰，在攪拌過程的最後3分鐘，加入鹽繼續攪拌完成。
3/ 完成攪拌後，加入已放涼的蘑菇以慢速拌勻。
4/ 把部份麵糰置冰箱，準備供碾平作麵包面層。
5/ 麵糰初步發酵完成後，放在已灑粗粒小麥粉的枱上，秤重量，搓圓，讓其鬆弛。
6/ 碾平辣味麵糰，厚約3毫米，用圓紙片作模，把麵糰切割成相同直徑。
7/ 置上已灑麵粉的布上，造成圓麵包。
8/ 在包面掃上清水，放進反轉麵糰置在碟上。
9/ 當發酵完成，翻轉麵包落於已灑麵粉的木板上。
10/ 放入蒸氣，直接落入烘焙石烘焗。當麵包開始轉色，開啟排氣活門。

1. Wash and cut the fresh mushrooms into thick slices. Cook them with olive oil and salt.
2. Knead the dough; add salt 3 minutes prior kneading ends.
3. Once the kneading is done, add the cooled mushrooms in slow speed.
4. Keep a part of dough in the fridge to later roll the top part of the loaves.
5. After the bulk fermentation, place the dough on a table dusted with fine semolina and weigh the dough. Gently pre-shape round loaves and allow resting.
6. Roll the chilled dough at 3mm thickness and cut discs of the same diameter as your loaves.
7. Place the discs on a cloth dusted with flour and shape your round loaves.
8. Brush water on the top of the loaves and place them upside- down on the discs.
9. Once fully proofed, flip the breads on a floured board.
10. Bake directly on your baking stone with steam. Open the exhaust once coloring starts.

蘋果酒浸葡萄乾芫荽煙燻煙肉包

CIDER RAISINS, SMOKED BACON & CORIANDER ROLLS

材料	INGREDIENTS	比例 RATIO
速成麵種	Poolish	
80 克 65 號麵粉	80 gm Flour type 65	
100 毫升蘋果酸	100 ml Apple cider	
2 克乾酵母	2 gm Dry yeast	
麵糰	Dough	
320 克 65 號麵粉	320 gm Flour type 65	100%
130 克 770 號黑麥麵粉	130 gm Rye flour type 770	
200 克成熟短暫發酵麵糰	200 gm Above ripe poolish	44%
1 克乾酵母	1 gm Dry yeast	-
12 克海鹽	12 gm Sea salt	2.7%
120 克酒浸葡萄乾	120 gm Raisins soaked in cider	26.7%
12 克芫荽籽碎	12 gm Coriander seeds crushed	2.7%
120 克炒香煙火腿	120 gm Smoked bacon sautéed	26.7%
350 毫升蘋果酒	350 ml Cider	77.8%
新鮮芫荽葉適量	Fresh coriander leaves	

建議 SUGGESTIONS

在醒發小麵卷的枱上灑一層薄薄的黑麥麵粉，這樣會給予較好的對比和味道。可加一個蘋果（金蘋果或加拿大蘋果）的粗丁粒混合麵糰，以添加果味和改善質地。這款麵包與煙肉、葡萄乾、芫荽和蘋果酒，十分匹配。

Use light rye flour to dust the table where you proof your rolls, it'll give you a better contrast and flavor. For an added fruity flavor and texture, add one apple (type Jonagold or Gala) in rough dice throughout the dough; very nice with bacon, raisins, coriander and cider.

64℃ A: 10 mins(S) 25℃ 2 hrs 50 gm
      B: 8 mins(F)

15 mins round rolls 2 hrs 230℃ 210℃ 25 mins

製法 METHOD

1 / 用拂蛋器拌勻速成麵種材料，置室溫下發酵4小時。
2 / 把葡萄乾完全浸泡於足夠的蘋果酒內4小時。
3 / 把煙肉切成長條，炒至深褐色。
4 / 混合所有材料，除了鹽、酒浸葡萄、煙肉和芫荽葉外，搓揉麵糰，在攪拌過程的最後3分鐘，加入鹽攪拌至完成。
5 / 麵糰攪拌一旦完成，以慢速拌入酒浸葡萄和煙肉。
6 / 鬆弛完成後，量小麵卷的重量，造型，放在木板上鬆弛。
7 / 做成圓形，鈒出1/3小麵糰，掃點清水。
8 / 貼上新鮮芫荽葉，反轉麵糰，置在已灑上麵粉的工作枱上發酵。
9 / 當發酵過程完成，反轉麵糰，放入蒸氣於焗爐內，烘焙至金黃色。

1. Prepare the poolish by mixing the ingredients with a whisk and keep it at room temperature to ferment for 4 hours.
2. Soak your raisins in enough cider to be submerged for the same amount of time.
3. Cut your smoked bacon in long strip and sautéed it dark brown.
4. Mix and knead your dough with all the ingredients except salt, raisins, bacon and coriander leaves. Add the salt 3 minutes prior kneading ends.
5. Once the dough is kneaded, add the raisins and bacon in slow speed.
6. Once rested, weigh the rolls, pre-shape and allow resting on the bench.
7. Shape round rolls, cut a ring through 1/3 of the roll. Brush a little water.
8. Stick a leaf of fresh coriander, and place them up-side-down on a floured surface for proofing.
9. Once proofed, flip them over and bake them with steam until golden brown.

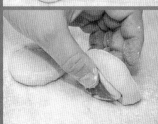

/ 蘋果酒浸葡萄乾芫荽煙燻煙肉包

紅甜椒瑞士芝士卷
RED BELL PEPPER & BAGNES CHEESE ROLLS

材料	INGREDIENTS	比例RATIO
380 克 65 號麵粉	380 gm Flour type 65	100%
75 克 黑麥麵粉	75 gm Rye flour	19.74%
6 克 乾酵母	6 gm Dry yeast	1.32%
12 克 鹽	12 gm Salt	2.64%
280 毫升 清水	280 ml Water	61.54%
115 克 瑞士巴谷山芝士	115 gm Bagnes cheese	25.27%
65 克 紅燈籠椒	65 gm Red bell pepper	14.29%
麵粉適量（灑面）	Dusting flour	

建議 SUGGESTIONS

瑞士巴谷山芝士源自瑞士瓦萊州。它被公認為最佳的Racletle芝士。這半硬芝士放麵包內烘焙或配已焗燈籠椒是非常完美組合。要是沒有這半硬芝士，你可用瑞士巴谷山芝士代替。芝士在烘焙時很快變棕褐色，所以在完成烘焙前10分鐘，你可以半開爐門。剛出的麵包上有正在熔化的芝士，需趁熱享用。它與芝麻菜沙律和成熟的車厘茄，十分匹配。

The Bagnes cheese from the Valais region in Switzerland. It is recognized as the finest cheese for Raclette. This type of semi-hard cheese is perfect to bake in bread and pairs nicely with roasted bell pepper. If not available, you can substitute Bagnes cheese with other semi-hard cheeses. Cheese will take a fast brown coloring while baking; therefore, at about 10 minutes from the end of baking, you might to keep your oven's door ajar. Freshly baked and served warm, this bread and its melting pieces of cheese will be the perfect fit for an arugula salad with ripe cherry tomatoes.

62℃	A: 10 mins(S) B: 5 mins(F)	25℃	1 hr	40 gm

15 mins	rolls	45 mins	235℃	210℃	30 mins

製法 METHOD

1/ 把紅燈籠椒放在明火爐燒至外皮焦黑。燒焦後，用保鮮紙包裹紅椒，停放15分鐘，這時外皮很容易去掉（不要浸水，因為水會浸淡紅椒的味道），然後去籽，切成約2至3厘米的小塊。

2/ 瑞士巴谷山芝士切成1厘米丁粒。

3/ 搓揉麵糰，在攪拌過程的最後3分鐘，加入鹽，繼續攪拌至完成。麵糰搓揉完成，以慢速加入芝士粒和紅燈籠椒塊。

4/ 初步發酵麵糰完成後，秤重量，造型。

5/ 在小麵糰底部掃點油，最後造型。

6/ 發酵後，把塗了油的麵糰翻轉向下，然放在已灑麵粉的發酵布上發酵。

7/ 發酵完成時，翻轉小麵糰，放入少量蒸氣，然後烘焙，待麵包轉色，立即開啟排氣活門。

1. Prepare the bell peppers; on a flame stove, burn until the skin in black color. Once fully burnt, wrap the pepper in a plastic film and let it rest for 15 minutes. You can then peel the pepper easily without using water bath. Water would reduce the flavors of the bell pepper. Then, empty the middle and cut the pepper in rough pieces of about 2 to 3 cm.

2. Dice the Bagnes cheese into cubes of about 1 cm.

3. Knead the dough and add salt 3 minutes prior the end of mixing. Once the kneading is done, add the diced cheese and bell pepper in slow speed.

4. After the bulk time, weigh the dough and pre-shape the rolls.

5. Brush a little oil on the bottom of the roll and give the final roll shape to the dough.

6. Proof with the oiled surface facing downward, on a floured cloth.

7. Once fully proofed, turn the rolls over and bake with a little steam. Open the steam exhaust once coloration starts.

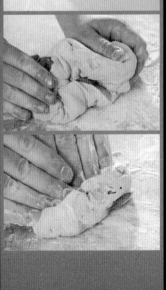

油浸蒜頭牛至扁麵包
Confit Garlic & Oregano Schiacciata

63℃ | A: 10 mins(S) / B: 6 mins(F) | 27℃ | 45 mins | 350 gm

15 mins | flat bread | 30 mins | 225℃ | 210℃ | 40 mins

建議 SUGGESTIONS

油浸烤蒜頭，在小煎鍋中燒熱橄欖油，慢火煮鮮蒜頭至軟身和完全熟透。

To prepare garlic confit, prepare a cooking pot with olive oil and simmer at low temperature the fresh garlic knob until soft and cooked through.

材料	INGREDIENTS	比例RATIO
麵種	**Starter**	
250 克 45 號麵粉	250 gm Flour type 45	100%
12 克 乾酵母	12 gm Dry yeast	4.8%
240 毫升 清水	240 ml Water	96%
麵糰	**Dough**	
400 克 45 號麵粉	400 gm Flour type 45	100%
100 克 雞蛋	100 gm Egg	25%
10 克 白砂糖	10 gm White sugar	2.5%
85 毫升 橄欖油	85 ml Olive oil	21.25%
12 克 鹽	12 gm Salt	3%
新鮮牛至香草適量	SQ Fresh oregano	
油浸烤蒜頭適量	SQ Confit garlic knob	
海鹽適量	SQ Sea salt	
橄欖油適量（掃面）	SQ Olive oil for brushing	
麵粉適量（灑面）	Dusting flour	

製法 METHOD

1/ 麵種材料用拂蛋器拌勻，置和暖地方下貯放2小時。
2/ 搓揉麵糰，加入麵種和一些牛至香草拌勻。
3/ 在攪拌過程的最後3分鐘，放入鹽續攪拌至完成。
4/ 當麵糰鬆弛完成，造成圓平狀。
5/ 放入已鋪烘焙紙的焗盤上，發酵麵糰。
6/ 醒發完成時，掃上橄欖油和用手指於麵糰戳些小孔。
7/ 把油浸烤蒜頭壓入麵糰內，再灑上海鹽和牛至香草。
8/ 不用放蒸氣於爐內，但需開啟排氣活門，烘焙至金黃色。

1. Mix the starter using a whisk and keep in a warm place for 2 hours.
2. Knead the dough, adding the starter and some oregano in the dough.
3. Add the salt 3 minutes prior kneading ends.
4. After resting time, shape the bread into round flat breads.
5. Place on a tray lined with baking paper and proof.
6. Once proofed, brush with olive oil and make holes with your fingers.
7. Push garlic confit knobs in the dough and sprinkle with sea salt and oregano.
8. Bake until golden brown, without steam and with exhaust opened.

香脆松子羅勒卷
CRISPY PINE NUTS BASIL ROLLS

63℃   A: 8 mins(S) / B: 8 mins(F)   24℃   30 mins   45 gm

no rest time   snail rolls   45 mins   210℃   190℃   25 mins

建議 SUGGESTIONS

麵糰的顏色平均，待烘焙時間已過了 3/4，脫去環模。把此麵糰卷發酵，技巧是要確保置放的溫度不太高，否則牛油會熔化，失去達至有層次感的效果。

建議 SUGGESTIONS

To obtain an even coloration of your roll, remove the ring after 3/4 of the baking time is past. When proofing this roll, ensure that the temperature of the place is not too high or else your butter will melt and the layering effect will be lost.

材料	INGREDIENTS	比例 RATIO
250 克 65 號麵粉	250 gm Flour type 65	100%
8 克 乾酵母	8 gm Dry yeast	3.2%
5 克 白砂糖	5 gm White sugar	2%
6 克 鹽	6 gm Salt	2.4%
50 克 牛油	50 gm Butter	20%
150 毫升 牛奶	150 ml Milk	60%
60 克 牛油（夾心）	60 gm Butter for folding	24%
松子仁適量（烘焙）	SQ Toasted pine nuts	
新鮮羅勒適量	SQ Fresh basil	
海鹽適量	Sea salt	
麵粉適量（灑面）	Dusting flour	

製法 METHOD

1/ 攪拌所有材料，並在攪拌過程的最後3分鐘，加入鹽續攪拌至完成。
2/ 把麵糰壓成方形，厚約2厘米，置冰箱待3小時。
3/ 夾心牛油放在兩片焗餅紙的中央碾平，做出纖薄牛油片。
4/ 放在麵糰的中央，然後把麵糰摺疊牛油，做法如牛角包。
5/ 麵糰捲起2厘米，再按麵糰分成3份，覆摺成3層。
6/ 捲起麵糰2厘米，重複覆摺2次。
7/ 包起麵糰，置冰箱1小時。
8/ 把麵糰碾平約3毫米厚，掃點清水。
9/ 灑上已烘焙的松子仁和新鮮羅勒碎，捲成長螺紋狀。
10/ 麵糰鬆弛，切成40克一份，放在已噴油的模具發酵。
11/ 灑上海鹽片，放蒸氣於焗爐，關閉排氣活門才開始烘焙。當麵糰轉色，立即開啟排氣活門。

1. Knead the dough with all ingredients; add the salt 3 minutes prior kneading ends.
2. Press the dough into a flat square of about 2 cm thickness and refrigerate for 3 hours.
3. Roll your folding butter between 2 pieces of baking paper to obtain a thin butter sheet.
4. Place the butter in the middle of the dough; fold the dough to cover the butter, like making croissant dough.
5. Roll at 2 cm and visually divide the dough in 3 parts and fold into 3 tiers.
6. Roll again the dough at 2 cm and repeat the folding 2 more times.
7. Wrap the dough and keep it in the fridge for 1 hour.
8. Roll the dough at 3 mm thickness and brush with a little water.
9. Sprinkle toasted pine nuts and fresh basil chiffonade; roll the dough into a long snail.
10. Relax the dough and cut into 40 gm portion. Place the dough into greased moulds and proof.
11. Sprinkle a few sea salt flakes and bake with steam and closed exhaust. Open the exhaust once coloring starts.

ALMOST
UNTOUCHABLE
巧味包

62℃   A: 10 mins(S)   24℃   none   450 gm
       B: 2 mins(F)

none   round   30 mins   215℃   195℃   45 mins

材料	INGREDIENTS	比例 RATIO
200 克 55 號麵粉	200 gm Flour type 55	100%
300 克 全麥麵粉	300 gm Whole wheat flour	
12 克 梳打粉	12 gm Baking soda	2.4%
10 克 鹽	10 gm Salt	2%
300 毫升 鮮牛奶	300 ml Fresh milk	60%
150 克 純味乳酪	150 gm Plain yogurt	30%
麵粉適量（灑面）	Dusting flour	

製法 METHOD

1. 把麵糰輕輕搓揉，只需用平常力度的七成便可。
2. 把麵糰搓圓，放在木製的發酵籃內，此法可令麵糰印出花紋。
3. 放置一旁，讓麵糰鬆弛30分鐘。
4. 把發酵木籃翻轉，倒出麵糰，然後在麵糰上刻"十"字紋。
5. 不用放蒸氣於焗爐內，關閉排氣活門，然後烘焙，待麵包轉色，開啟排氣活門。

1. Knead the dough lightly, not as much as regular dough, but until 70% of the kneading.
2. Shape the boule and place them in fermenting wooden basket to give them shape and design of flour.
3. Let it rest for 30 minutes.
4. Flip it out of the wooden basket and score a cross on the bread.
5. Bake without steam and closed exhaust. Open the exhaust once coloring starts.

建議 SUGGESTIONS

製造愛爾蘭梳打包非常快捷簡單。事實上，它屬於速成麵包的類別，因發酵方法由梳打粉的化學反應而成，並沒有經過正確的發酵程序。我享用這包時會伴以鹹味牛油和美味的愛爾蘭農場芝士。當我在愛爾蘭工作時，就完全沉浸在當地品種繁多的優質芝士裏。

The making of Irish soda bread is quick, as a matter of fact; it falls in the category of quick breads; meaning that no fermentation process is involved since the leavening method is chemical with the baking soda. I enjoy this bread with salty butter and the great Irish farm cheeses. During my time working in Ireland, I was positively submerged by the variety and quality of Irish cheeses.

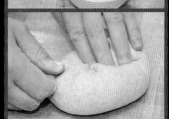

種籽麵包
SEEDED OVAL

材料	INGREDIENTS	比例RATIO
300 克 65 號麵粉	300 gm Flour type 65	
65 克 全麥麵粉	65 gm Whole wheat flour	100%
65 克 黑麥麵粉	65 gm Rye flour	
150 克 麵種	150 gm Fermented dough	34.88%
5 克 乾酵母	5 gm Dry yeast	1.16%
75 克 雜種籽	75 gm Seeds mix	17.44%
15 克 鹽	15 gm Salt	3.49%
360 毫升 清水	360 ml Water	83.72%
麵粉適量（灑面）	SQ Dusting flour	

建議 SUGGESTIONS

混合種籽可由葵花籽、芝麻、壓碎的小麥、亞麻籽、南瓜籽或燕麥粒等組成，按個人喜好便可。

麵種是用早一天用剩存放在冰箱中的麵糰。如果沒有在手，可以做一塊小麵糰。用65號麵粉約150克、100毫升清水、2或3克酵母，搓揉成糰，放在室溫下發酵2小時，放進冰箱就可在翌日使用。

Seeds mix can be made of sunflower seeds, sesame, cracked wheat, flaxseed, pumpkin seeds or oat kernel and so on, depending on your taste.

The fermented dough should be leftover dough from the day before stored in the fridge. If you don't have it on hand, you can produce small dough. To produce the fermented dough, you can use 150 gm of flour type 65 and 100 ml of water with 2 or 3 grams of yeast, knead it and leave it to ferment for 2 hours at room temperature; keep it in the fridge until the next day.

巧味包 Almost untouchable

62℃　　　　　A: 10 mins(S)　　　25℃　　　　　2 hrs　　　　　350 gm
　　　　　　　B: 8 mins(F)

20 mins　　　　　oval　　　　　1 hr　　　　　225℃　　　　　200℃　　　　　45 mins

製法 METHOD

1/ 搓揉麵糰，在攪拌過程的最後3分鐘，加入鹽續攪拌至完成。
2/ 麵糰搓揉完成，以慢速拌入穀類種籽。
3/ 麵糰鬆弛後，秤重量，造型呈橄欖狀，置木板上鬆弛。
4/ 再造成橄欖形，放在已灑麵粉的布上發酵。
5/ 發酵完成，在麵糰上面切一刀，直接放在烘焙石烘焗。
6/ 放蒸氣於焗爐內，關閉排氣活門，然後烘焙，待麵包轉色，立即開啟排氣活門。

1. Knead the dough and add the salt 3 minutes before kneading ends.
2. Add the grains in slow speed once the dough is fully kneaded.
3. After the resting time, weigh the dough and pre-shape ovals; rest it on the bench.
4. Shape the ovals and place them on a floured fermenting cloth.
5. Once fully proofed, give one large cut on top of the loaf and bake directly on baking stone.
6. Bake with steam and closed exhaust. Open the exhaust once coloring starts.

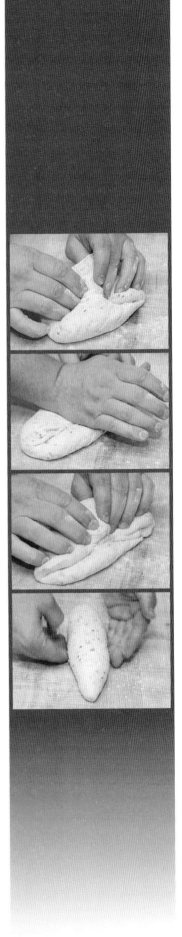

蕎麥枕頭麵包
BUCKWHEAT PILLOW BREAD

材料	INGREDIENTS	比例 RATIO
250 克 65 號麵粉	250 gm Flour type 65	
75 克 全麥麵粉	75 gm Whole wheat flour	100%
50 克 蕎麥麵粉	50 gm Buckwheat flour	
150 克 發酵麵糰	150 gm Fermented dough	40%
3 克 乾酵母	3 gm Dry yeast	0.8%
20 克 小麥胚芽	20 gm Wheat germs	5.33%
15 克 鹽	15 gm Salt	4%
350 毫升 清水	350 ml Water	93.33%
麵粉適量（灑面）	SQ Dusting flour	

建議 SUGGESTIONS

用軟身的紙卡或紙張刻出模版，紙質必須與已發酵麵糰的細緻表面吻合，並確保模版表面乾燥，不會黏住麵糰。

Cut a stencil in a soft cardboard or paper that can adapt to the delicate surface of the proofed bread. Ensure the surface to be dry enough not to have the stencil sticking to the dough.

64℃	A: 8 mins(S) B: 8 mins(F)	26℃	18 hrs at 5℃	800 gm

3 hrs	rectangle load	1 hr	225℃	200℃	50 mins

製法 METHOD

1 / 搓揉麵糰，在攪拌過程的最後3分鐘，加鹽續攪拌至完成。
2 / 經過長時間發酵後，讓麵糰置室溫下待3小時。
3 / 碾平麵糰，覆摺成3層，如牛角包做法。
4 / 鬆弛麵糰10分鐘，再次碾平，約4厘米厚。
5 / 切割成長方形麵糰，放在已灑麵粉的發酵布上發酵。
6 / 準備模版，待麵糰醒發後放上，再灑上黑麥麵粉。
7 / 放入蒸氣入烤爐，關閉排氣活門，然後烘焙，待麵包開始轉色，立即開啟排氣活門。

1. Knead the dough and add the salt 3 minutes before kneading ends.
2. After the long bulk fermentation, let the dough sit at room temperature for 3 hours.
3. Flatten the dough and fold it in 3 layers once, like for croissant dough.
4. Allow resting for 10 minutes and flatten it again to about 4 cm in thickness.
5. Cut the rectangles of dough and place on a floured cloth to proof.
6. Prepare your stencil and once proof, place the stencil on the bread and dust with rye four.
7. Bake with steam and closed exhaust. Open the exhaust once coloring starts.

洛神花黑麥麵包
Light Roselle Rye Boule

65℃ | A: 10 mins(S) B: 6 mins(F) | 25℃ | 2 hr | 400 gm

15 mins | round | 1.5 hrs | 235℃ | 215℃ | 40 mins

建議 SUGGESTIONS

帶酸和似有若無的洛神花花香，與黑麥的大地氣息相當搭配。它能帶出不平凡又令人愉快的味道，是黑麥麵包的新搭擋。

The acidic and subtle blossom fragrance of Roselle fits very well the earthy flavors of rye. It brings a very pleasant uncommon flavor and new twist to rye bread.

材料　INGREDIENTS　比例 RATIO

材料	INGREDIENTS	比例 RATIO
300 克 65 號麵粉	300 gm Flour type 65	100%
150 克 黑麥麵粉	150 gm Rye flour	
2 克 乾酵母	2 gm Dry yeast	0.44%
100 克 麵種	100 gm Fermented dough	22.22%
12 克 海鹽	12 gm Sea salt	2.67%
100 克 洛神花果醬	100 gm Roselle jam	22.22%
375 毫升 清水	375 ml Water	83.33%
麵粉適量（灑面）	SQ Dusting flour	

製法 METHOD

1. 搓揉麵糰，在攪拌過程的最後3分鐘，加入鹽續攪拌至完成。
2. 麵糰攪拌完成，以慢速拌入洛神花果醬。
3. 鬆弛麵糰，做圓麵糰，放在木板上鬆弛。
4. 再造型，把麵糰翻轉在木質發酵籃。
5. 讓其完全發酵，翻轉麵包，直接放在烘焙石上烘焗。
6. 放入蒸氣於焗爐中，關閉排氣活門，然後烘焙，待麵包開始轉色，立即開啟排氣活門。

1. Knead the dough and add the salt 3 minutes before kneading ends.
2. Add the Roselle jam after the kneading is done, in slow speed.
3. After the resting time, pre shape in round loaf and let it rest on the bench.
4. Shape the loaves and place them up-side-down in wooden proofing basket.
5. Once fully proofed, turn the loaves over and bake directly on baking stone.
6. Bake with steam and closed exhaust. Open the exhaust once coloring starts.

孜然波爾多皇冠包
Caraway Bordelaise Crown

64℃ | A: 10 mins(S) B: 8 mins(F) | 25℃ | 1 hr | 350 gm total

30 mins | rolls in crown | 1 hr | 235℃ | 215℃ | 35 mins

建議 SUGGESTIONS

傳統的孜然黑麥麵包，皇冠造型是非常優美和吸引。不要過度用力拉扯小卷底下的三角麵皮，因這樣做會在烘焙時令它變形。取而代之，可切割大塊的圓麵糰，讓其有多些覆蓋面。

Classic caraway and rye bread, the shape of the Bordelaise crown is nice and attractive. Try not to pull the triangle too hard under each roll, as they will end up deforming during the baking. Instead, cut a larger disc of dough to allow plenty of coverage.

材料	INGREDIENTS	比例RATIO
150 克 65 號麵粉	150 gm Flour type 65	100%
150 克 770 號黑麥麵粉	150 gm Rye flour type 770	
240 毫升 清水	240 ml Water	80%
4 克 乾酵母	4 gm Dry yeast	1.33%
10 克 麥芽糖液	10 gm Liquid malt	3.33%
2 克 孜然（葛縷子）	2 gm Caraway seeds	0.67%
8 克 鹽	8 gm Salt	2.67%
麵粉適量（灑面）	SQ Dusting flour	

製法 METHOD

1/ 搓揉麵糰，在攪拌過程的最後3分鐘，加入鹽續攪拌至完成。

2/ 麵糰發酵完成，分割麵糰，做成小卷。可預留部份麵糰，存放在冰箱半小時，取出後捲起用作麵包中央。

3/ 準備一個發酵籃，將一隻碗反轉放於籃中央，用布蓋上。

4/ 取出冷凍麵糰，碾平成圓片狀，大小剛夠蓋住反轉的碗。將麵塊鋪入籃前，在圓形麵塊中央對角切出三角形。

5/ 將麵塊鋪在碗上，並整理小卷的造型，將小卷反轉放在圓麵塊上。

6/ 在麵包底部掃點清水，覆摺每個小卷底下的三角的尖端。

7/ 待其完全醒發，把發酵籃翻轉，把整組麵糰直接放在烘焙石上。

8/ 放入蒸氣，關閉排氣活門，然後烘焙，待麵包開始轉色，立即開啟排氣活門。

1. Knead the dough and add the salt 3 minutes before kneading ends.

2. After the bulk time, divide the dough and pre-shape the rolls. Keep a part of dough in the fridge for half an hour to roll the central part.

3. Prepare the proofing basket with an up-side-down bowl in the middle, covered with a cloth.

4. Roll the chilled dough in a disc that covers the bowl surface; before leaning the dough in the basket, cut the triangles on the middle of the disc.

5. Lean your disc on the bowl and shape the rolls. Place them up-side-down on the disc of dough.

6. Brush a little water on the base of the rolls and fold the tip of each triangle under the rolls.

7. Once proofed, flip the basket over and bake directly on baking stone.

8. Bake with steam and closed exhaust. Open the exhaust once coloring starts.

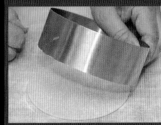

法式長麵包
FRENCH BAGUETTE

建議 SUGGESTIONS

使用速成麵種及經中等長度的發酵時間是一種令麵包味道豐厚和外層香脆的好方法。製作速成麵種，只要用室溫的麵粉和清水便可；使用過熱的水，會令麵種過度發酵，不能讓麵包達至預期效果。速成麵種可以長時間貯存在溫度5℃而發酵，較長時間發酵可令其味道更好和可延長保存期。

Using a poolish is a great way to produce flavorful and crusty bread over a medium time of fermentation. When making the poolish, use room temperature flour and water. Using an excessively warm water will eventually result in an over proofed poolish. It will also not produce the expected fermentation in the final dough. The poolish can be made over longer period of time kept to ferment at 5°C. Longer time of fermentation means better flavors and conservation.

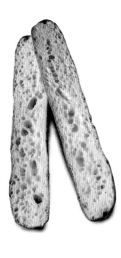

64℃   A: 10 mins(S) / B: 8 mins(F)   26℃   1.5 hr   350 gm

15 mins   baguette   1 hr   235℃   215℃   35 mins

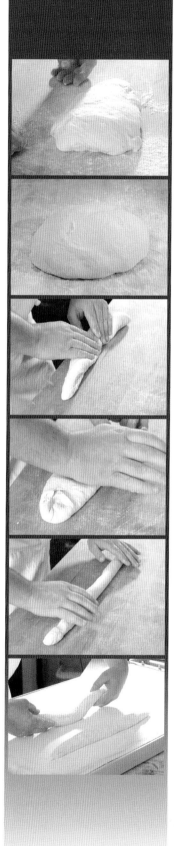

材料	INGREDIENTS	比例RATIO
原始麵種	**Starter**	
90 克 65 號麵粉	**90 gm Flour type 65**	
90 毫升 清水	**90 ml Water**	
3 克 乾酵母	**3 gm Dry yeast**	
麵糰	**Dough**	
180 克 65 號麵粉	**180 gm Flour type 65**	**100%**
110 毫升 清水	**110 ml Water**	**61.11%**
2 克 乾酵母	**2 gm Dry yeast**	**1.11%**
180 克 麵種	**180 gm Starter (all of above)**	**100%**
6 克 鹽	**6 gm Salt**	**3.33%**
麵粉適量（灑面）	**Dusting flour**	

製法 METHOD

1/ 混合麵種材料成幼滑麵糰，置室溫下發酵5小時。
2/ 搓揉麵糰，在攪拌過程的最後3分鐘，加鹽續攪拌至完成。
3/ 鬆弛麵糰，秤重量，造型成長棒形，放在木板上鬆弛。
4/ 當鬆弛完成，再把法包造型，在造型後灑少許麵粉發酵。
5/ 發酵完成時，用刀在法包上刻紋4-5下，置在烘焙石直接烘焗。
6/ 放入蒸氣，關閉排氣活門，然後烘焙，待麵包開始轉色，立即開啟排氣活門。

1. Mix the starter in smooth dough and let it ferment at room temperature for 5 hours.
2. Knead the dough and add the salt 3 minutes before kneading ends.
3. Allow resting and weigh the dough. Pre-shape baguette and let rest on bench.
4. Once rested, shape the baguette and end the shaping with a slight touch of flour.
5. Once proofed, score the baguette 4 to 5 times and bake directly on the baking stone.
6. Bake with steam and closed exhaust. Open the exhaust once coloring starts.

鄉村麵包棍
FABRICE COUNTRY STICK

建議 SUGGESTIONS

這食譜發酵過程用上24小時有別於其他方法。這技術可讓麵糰和麵筋非常流質狀，隨時間結在一起。麵糰將仍然柔軟，但烘焗無論麵包外層或內層，均可助製作出很出色的質感、麵包屑和脆碎層。這種麵包的保存期可較長。

這麵包現已普及於歐洲，但發展初期，我和好友 —— 費比斯和工作團隊，花上無數小時研發出一個較長時間而作做出的麵糰方法，繼續發展出這新款的品種，給予人驚喜的農村麵包。

The fermentation process technique of more than 24 hours allows this very liquid dough's gluten to coagulate over time. The dough will still be soft, but when baking it will create a pleasant texture, crumb and crust. This bread has a fairly long time of conservation.

At the early stage of development of this now popular bread in Europe, my good friend Fabrice and our team worked endless hours to develop this version of the loaf into a more matured dough method, giving you this fantastic rustic bread.

54℃   A: 8 mins(S)   22℃   1st 24 hrs at 5℃   380 gm
B: 8 mins(F)   2nd 3 hrs at 25℃

15 hrs   twisted baguette   40 mins   245℃   220℃   35 mins

材料	INGREDIENTS	比例 RATIO
600 克 65 號麵粉	600 gm Flour type 65	
25 克 全麥麵粉	25 gm Whole wheat flour	100%
35 克 黑麥麵粉	35 gm Rye flour	
100 克 發酵麵糰	100 gm Fermented dough	15.15%
3 克 乾酵母	3 gm Dry yeast	0.45%
15 克 小麥胚芽	15 gm Wheat germs	2.27%
15 克 鹽	15 gm Salt	2.27%
550 毫升 清水	550 ml Water	83.33%
麵粉適量（灑面）	SQ Dusting flour	

製法 METHOD

1 / 搓揉麵糰，在攪拌過程的最後3分鐘，加入鹽續攪拌至完成。麵糰將非常柔軟，有流質的感覺。
2 / 把麵糰保存在大膠盒內24小時，它會變成大3倍。
3 / 繼續待在溫度25℃的地方發酵3小時，然後放在灑滿大量麵粉的枱上。
4 / 小心翼翼地把量麵糰量重，再搓成長條形，在麵包粉上扭扣在一起。
5 / 放在發酵布待40分鐘。
6 / 確保麵包沒有黏貼在布上，然後直接放在烘焙石上烘焙。
7 / 放入蒸氣於焗爐內，關閉排氣活門，然後烘焙，待麵包開始轉色，立即開啟排氣活門。

1. Knead the dough and add the salt 3 minutes before kneading ends. The dough will be very soft with a liquid texture.
2. Keep the dough in large plastic containers; the dough will triple its size in the next 24 hours.
3. Keep the dough in bulk at 25°C for 3 hours; place the dough on the heavily floured table.
4. Carefully, weigh the dough into long shape pieces and twist them in bread flour.
5. Place on the fermenting cloth for 40 minutes.
6. Make sure the breads don't stick to the cloth and bake directly on the stone.
7. Bake with steam and closed exhaust. Open the exhaust once coloring starts.

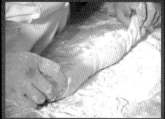

巧味包 Almost untouchable

斯佩爾特小麥農夫麵包
SPELT FARMER BREAD

材料	INGREDIENTS	比例RATIO
300 克 65 號麵粉	300 gm Flour type 65	
70 克 全斯佩爾特小麥麵粉	70 gm Whole spelt flour	100%
70 克 770 號黑麥麵粉	70 gm Rye flour type 770	
5 克 乾酵母	5 gm Dry yeast	1.14%
15 克 小麥胚芽	15 gm Wheat germs	3.41%
10 克 鹽	10 gm Salt	2.27%
350 毫升 清水	350 ml Water	79.55%
麵粉適量（灑面）	SQ Dusting flour	

建議 SUGGESTIONS

鄉村麵包有其獨特的小麥味道，斯佩爾特小麥與芝士和酒非常搭配。這麵包經烘烤後帶出小麥裏的果仁味道，因而特別與芝士特別匹配。如要獲得濃郁的小麥味道，可把部份麵包粉以全麥麵粉取代，但質感密度會相對地增加。確切地，如麵粉份量多些粗糙感，當發酵至麵筋網絡時便容易受到影響。如希望把「帽子」貼在麵包頂部，把麵糰平坦的部份覆疊在頂部前，先掃上清水，否則麵包部便會在烘焗後上升。

The country bread has a typical taste of wheat and spelt and goes well with cheese and wine. Toasting the country bread brings out the nutty flavor of wheat and makes it particularly pleasant with cheeses. To obtain a stronger wheat flavor, you can substitute some bread flour and replace it with whole wheat flour, but the texture will increase in density. Indeed, with a more coarse flour mix, there will be less small particles of starch susceptible to ferment into a broad network of gluten. If you wish to have the "hat" of the bread sticking to the bread, brush some water before folding the flatten part over the top, else, the top part will rise.

* 封面上的麵包是斯佩爾特小麥農夫麵包的另一演繹。
  Cover bread is another presentation of spelt farmer bread.

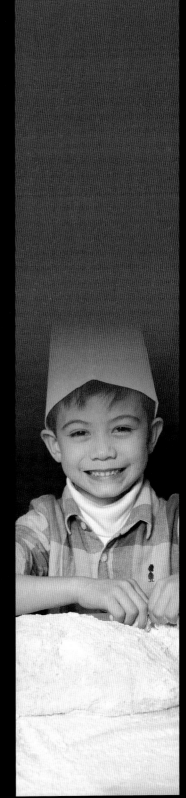

65℃	A: 10 mins(S) B: 6 mins(F)	25℃	2 hr	450 gm

| 15 mins | round | 1.5 hrs | 235℃ | 215℃ | 45 mins |

製法 METHOD

1/ 搓揉麵糰，在攪拌過程的最後3分鐘，加入鹽續攪拌至完成。
2/ 麵糰攪拌完成，造型，呈圓形狀，鬆弛。
3/ 再造型，把一面碾薄，覆上頂面，置麵包粉在工作枱上，輕輕繼續碾平麵糰。
4/ 反轉麵糰置放在發酵布上。
5/ 完成發酵後，翻轉麵包，直接放在烘焙石上烘焗。
6/ 放入蒸氣於爐內，關閉排氣活門，然後烘焙，待麵包開始轉色，立即開啟排氣活門。

1. Knead the dough and add the salt 3 minutes before kneading ends.
2. After the resting time, pre-shape in round loaf and allow resting.
3. Shape the loaves and roll thinly one side to fold it over the top. Roll slightly in bread flour.
4. Place on the fermenting cloth up-side-down.
5. Once fully proofed, turn the loaves over and bake directly on baking stone.
6. Bake with steam and closed exhaust. Open the exhaust once coloring starts.

/ 斯佩爾特小麥農夫麵包

全麥吐司
WHOLE WHEAT PAIN DE MIE

巧味包 Almost untouchable

材料	INGREDIENTS	比例RATIO
700 克 65 號麵粉	700 gm Flour type 65	
300 克 全麥麵粉	300 gm Whole wheat flour	100%
50 克 小麥皮	50 gm Wheat bran	
15 克 乾酵母	15 gm Dry yeast	1.43%
15 克 鹽	15 gm Salt	1.43%
100 克 牛油	100 gm Butter	9.52%
100 克 砂糖	100 gm Sugar	9.52%
100 毫升 鮮牛奶	100 ml Fresh milk	9.52%
600 毫升 清水	600 ml Water	57.14%
全麥麵粉適量（灑面）	SQ Whole wheat flour for dusting	
清水適量（掃面）	Water	

建議 SUGGESTIONS

添加小麥皮於全麥麵包可令切成薄片後帶出果仁味道，和更有口感，烘烤後加上果醬、牛油和任何英式早點搭配，便可作一頓豐富的早餐。

想麵包的質地緊密，可把麵糰分成4份，造成圓型，邊貼邊放於已塗油的模內。麵包烘烤後，最重要是確保麵糰造型時沒有氣泡，重要避免烘焙後出現大氣泡。

Adding some wheat bran to the whole wheat loaf will bring a nutty taste when the slice will be toasted for breakfast with jams and butter or with any English breakfast dishes.

To obtain a tighter texture in the loaf, it is possible to shape it divided in 4 parts of dough in round shape placed next to each other in the greased mould. In toast bread, it is important to shape the loaf without air bubble in order to avoid any large holes after baking.

54℃   A: 8 mins(S)   25℃   1 hr   450 gm
             B: 6 mins(F)

15 mins   rectangle loaf   2 hrs   215℃   190℃   40mins

製法 METHOD

1. 搓揉麵糰，在攪拌過程的最後3分鐘，加鹽續攪拌至完成。
2. 完成麵糰鬆弛後，把麵糰造型，放入已塗油的吐司模。
3. 發酵完成後，在麵包面掃點清水，灑上全麥麵粉。
4. 放入蒸氣於焗爐，關閉排氣活門，然後烘焙，待麵糰轉色，立即開啟排氣活門。
5. 烘焙完成後短時間脫模，放在涼凍架上貯放，否則吐司底部會因冷凝而變濕潤。

1. Knead the dough and add the salt 3 minutes before kneading ends.
2. After the resting times, shape the loaves and place them in greased loaf moulds.
3. Once fully proofed, brush the top of the loaf with water and sprinkle whole wheat flour.
4. Bake with steam and closed exhaust. Open the exhaust once coloring starts.
5. Unmould the baked loaves shortly after baking and place them on a cooling grid or else condensation will make the base of the bread damp.

巧味包 Almost untouchable

147

建議 SUGGESTIONS

你可以把麵包造型狀條或分成2至3細小條。甚至造成圓形，只要邊貼邊放在吐司模內便可。

額外添加海藻和檸檬會在麵包內可令麵包有細緻的效果，希望能給予你製做法式撻和三文治一些新靈感。

You can shape the loaf either in one piece or you can divide it in 2 or 3 pieces. Shape round loaves and place them next to each other in the toast mould.

The addition of seaweed and lemon gives this loaf a fine finish, giving you new ground for creative tartines and sandwiches.

海藻檸檬絲吐司
SEAWEED LEMON PAIN DE MIE

52℃	A: 12 mins(S) B: 6 mins(F)	24℃	1 hr	550 gm

30 mins	square toast	2 hrs	215℃	195℃	45 mins

材料 INGREDIENTS	比例RATIO
500 克 45 號麵粉　　500 gm Flour type 45	100%
25 克 雞蛋　　25 gm Eggs	5%
7 克 乾酵母　　7 gm Dry yeast	1.4%
7 克 鹽　　7 gm Salt	1.4%
50 克 牛油　　50 gm Butter	10%
50 克 砂糖　　50 gm Sugar	10%
50 毫升 鮮牛奶　　50 ml Fresh milk	10%
280 毫升 清水　　280 ml Water	56%
12 克 乾海藻　　12 gm Dried seaweed	2.4%
1 個檸檬皮　　1 pc Lemon zest	

製法 METHOD

1/ 搓揉麵糰，在攪拌過程的最後 3 分鐘，加鹽續攪拌至完成。
2/ 麵糰攪拌完成，以慢速拌入檸檬皮和海藻。
3/ 完成發酵，搓成長包形，放入已塗油的吐司模。
4/ 發酵完成時，放入蒸氣於焗爐，關閉排氣活門，然後烘焙，待麵糰轉色，立即開啟排氣活門。
5/ 烘焙完成後在短時間內脫模，放在涼凍架上貯放，否則吐司底部會因冷凝而變濕潤。

1. Knead the dough and add the salt 3 minutes before kneading ends.
2. Add the lemon zest and seaweed at the end of kneading in slow speed.
3. After the resting times, shape the long loaves and place them in the greased toast moulds.
4. Once fully proofed, bake with steam and closed exhaust; open it once the coloring starts.
5. Unmould the baked loaves shortly after the end of baking. Store them on a cooling grid or else condensation will make the base of the bread damp.

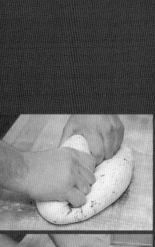

經典意大利軟包
CLASSIC CIABATTA

材料	INGREDIENTS	比例RATIO
75 克 45 號麵粉	75 gm Flour type 45	⎤ 100%
300 克 65 號麵粉	300 gm Flour type 65	⎦
350 毫升清水	350 ml Water	93.33%
5 克乾酵母	5 gm Dry yeast	1.33%
6 克鹽	6 gm Salt	1.6%
20 毫升橄欖油	20 ml Olive oil	5.33%
麵粉適量（灑面）	Dusting flour	

64℃   A: 10 mins(S)   26℃   24 hrs at 5℃   350 gm
        B: 8 mins(F)

10 mins   rectangle loaf   1 hr   235℃   215℃   35 mins

製法 METHOD

1/ 混合材料和攪拌成糰，放入連蓋的膠盒內，置冰箱貯藏。

2/ 待12小時後，取出麵糰摺覆，續保存在冰箱12小時。

3/ 注意麵糰將會非常柔軟，所以要使用足夠麵粉灑在枱上和膠盒內。

4/ 碾平麵糰約3厘米厚，鬆弛10分鐘，分切成350克重的長方形。

5/ 放入焗盤或發酵布，再次確定那裏有足夠麵粉。

6/ 當發酵完成，放蒸氣入焗爐，關閉排氣活門，然後烘焙，待麵包轉色，立即開啟排氣活門。

1. Mix and knead the dough; store it in a covered plastic container in the fridge.

2. After 12 hours, give the dough a fold. Keep it in the fridge for the remaining 12 hours.

3. Note that the dough will be very soft – use enough flour to dust tables and containers.

4. Flatten the dough to 3 cm thick; allow resting 10 minutes and cut rectangles of 350 gm.

5. Place on baking tray or fermenting cloth. Again, make sure there is enough flour dusted.

6. Once proofed, bake with steam and closed exhaust. Open the exhaust once coloring starts.

建議 SUGGESTIONS

意大利軟包是最古老的意式麵包。每個省份有其獨特款式 —— 有些麵包外層很柔軟；有些含脆香碎屑。可用牛奶替代清水，或可用全麥麵粉，令其變深色。麵包師亦會以香草、香料和其他特別材料作為做特色麵包的基礎。

品嚐意大利軟包，其中一種最佳食法是做三文治或可配一片優質帕爾馬芝士和初榨純橄欖油。把意大利軟包烤成吐司稱為帕尼洛三文治。軟包的主要特色是份量而質感幼細，入口輕軟。

Ciabatta is one of the most ancient Italian bread. Each region has its own version – some have soft crust and some have crispy crust. It is possible to make it with milk instead of water or to use whole wheat flour to make it darker. Bakers use the Ciabatta dough as a base for specialty bread, adding herbs, spices and other special ingredients.

Ciabatta is great for sandwiches or to accompany a nice piece of Parmesan cheese with some virgin olive oil for example. Toasted, the Ciabatta sandwich is known as panino. The key feature of the Ciabatta is its large and fine texture which makes it lighter to eat.

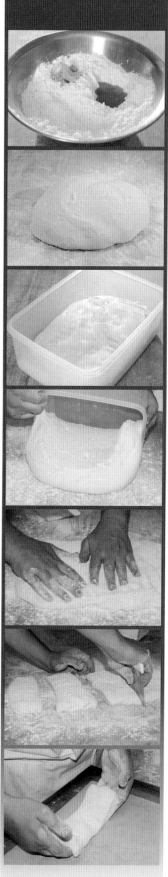

瑞士辮子麵包
PETITE TRESSE AU BEURRE

60℃ | A: 8 mins(S) B: 6 mins(F) | 24℃ | 30 mins | 250 gm

10 mins | braided | 40 mins | 200℃ | 190℃ | 35 mins

巧味包 Almost untouchable

材料	INGREDIENTS	比例 RATIO
440 克 45 號麵粉	440 gm Flour type 45	100%
200 毫升牛奶	200 ml Milk	45.45%
10 克乾酵母	10 gm Dry yeast	2.27%
40 克白砂糖	40 gm White sugar	9.09%
6 克鹽	6 gm Salt	1.36%
40 克雞蛋	40 gm Eggs	9.09%
60 克牛油	60 gm Butter	13.64%
雲呢拿香油適量	SQ Vanilla essential oil	
麵粉適量（灑面）	Dusting flour	
蛋液適量（掃面）	Egg wash	
75 克蜜餞薑粒	75gm Candied Ginger	

製法 METHOD

1 / 搓揉麵糰，在攪拌過程的最後3分鐘，加鹽續攪拌至完成。
2 / 麵糰鬆弛後，造成辮子形。
3 / 在焗盤鋪上矽膠紙。
4 / 當麵糰發酵後，掃上蛋液。
5 / 放入蒸氣入焗爐，烘焙至金黃色，待麵包開始轉色，立即開啟排氣活門。

1. Knead the dough and add the salt 3 minutes prior kneading ends.
2. After the resting time, shape the dough in braid shape.
3. Place on baking tray with silicon paper.
4. Once fully proofed, brush the loaves with egg wash.
5. Bake until golden brown with steam; open exhaust once coloring starts.

建議 SUGGESTIONS

由於含有高量酵母，這麵包的鬆弛時間較其他麵糰溫度也較凍，所以製作麵糰必須運作很快，否則麵包最終會出現大氣泡。完成製作這麵包後，可灑上少量芝麻作裝飾。

The resting times of this bread are shorter and the temperature of dough cooler due to the higher amount of yeast; the dough has to be worked quicker or else you will end up with large bubbles in the finished bread. Eventually finish it with some sesame seeds.

字母包
CRUMB

64℃　　　A: 10 mins(S)　　25℃　　45 mins　　variable
　　　　　B: 8 mins(F)

15 mins　　cut letter　　45 mins　　210℃　　200℃　　30 mins

材料	INGREDIENTS	比例RATIO
500 克 45 號麵粉	500 gm Flour type 45	100%
140 克牛油	140 gm Butter	28%
50 克白砂糖	50 gm Sugar	10%
10 克鹽	10 gm Salt	2%
15 克乾酵母	15 gm Dry yeast	3%
280 毫升鮮牛奶	280 ml Fresh milk	56%
1 個檸檬皮	1 pc Lemon zest	
130 克葡萄乾	130 gm Raisins	26%
蛋液適量（掃面）	SQ Egg wash	
麵粉適量（灑面）	Dusting flour	

製法 METHOD

1/ 搓揉麵糰，在攪拌過程的最後3分鐘，加入鹽續攪拌至完成。
2/ 麵糰攪拌完成，以慢速拌入葡萄乾和檸檬皮。
3/ 麵糰鬆弛後，把麵糰碾平一次，做出約2厘米厚。
4/ 放入已灑麵粉的焗盤，蓋上保鮮紙，貯放於冰箱1小時。
5/ 用大字模鈒出字形，放在已鋪焗餅紙的焗盤上。
6/ 讓其發酵，掃上一次蛋液，待乾，再次掃上蛋液。
7/ 放入蒸氣於焗爐，關閉排氣活間，然後烘焙至金黃色，待麵包開始轉色，立即開啟排氣活門。

1. Knead the dough and add the salt 3 minutes before kneading ends.
2. Add the raisins and lemon zest in slow speed once the kneading is done.
3. Once rested, fold the dough once and flatten it to about 2 cm in thickness.
4. Place on a floured tray, cover with plastic film and allow to rest in the fridge for 1 hour.
5. Cut the shapes using large letter cutters and place them on a tray with baking paper.
6. Allow proofing and brush egg wash once. Let it dry a little and apply egg wash once more.
7. Bake golden brown with steam and closed exhaust. Open the exhaust once coloring starts.

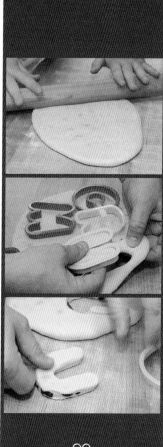

巧味包 Almost untouchable

建議 SUGGESTIONS

你可以把麵糰碾平一點，然後鈒成小字母，可做出迷你三文治或讓它浮在湯上。必定為兒童生日派對帶來驚喜。

讓麵糰置冰箱進行鬆弛，可在切割時令其造型輪廓鮮明，不會變形。掃蛋液兩次，可讓麵包帶光澤，但烘焙時要小心，因上顏色的速度會比平常快。

You can roll the dough thinner and cut smaller letters for example for mini sandwiches or to float on a soup. Great bread for kid's birthday parties!

Allowing the dough to rest in the fridge will give you sharp shapes and sharp cuts as the dough will not be deformed like it would be if cut right after the dough is flatten. Twice the egg wash is giving you a shiny look, but be careful at baking, it colors faster than usual.

辣根亞麻籽粗麥方塊包
Horseradish Flaxseed Semolina Squares

60℃　　　　A: 8 mins(S)　　　25℃　　　1 hr　　　50 gm
　　　　　　B: 6 mins(F)

15 mins　　square　　　30 mins　　215℃　　200℃　　25 mins

材料	INGREDIENTS	比例RATIO
500 克 優質粗麥麵粉	500 gm Fine semolina	100%
10 克 乾酵母	10 gm Dry yeast	2%
50 克 牛油	50 gm Butter	10%
12 克 鹽	12 gm Salt	2.4%
300 毫升 清水	300 ml Water	60%
50 克 辣根	50 gm Horseradish	10%
10 克 亞麻籽	10 gm Flaxseed	2%
粗麥麵粉適量（灑面）	SQ Semolina for dusting	

製法 METHOD

1/ 搓揉麵糰，在攪拌過程的最後3分鐘，加鹽續攪拌至完成。
2/ 麵糰攪拌完成後，以慢速拌入辣根和亞麻籽。
3/ 完成初步發酵麵糰，碾平麵糰約2厘米厚，蓋上保鮮紙，放冰箱冷凍1小時。
4/ 把厚身的麵糰再切成約5厘米的方塊，置於焗盤上。
5/ 發酵完成時，放入蒸氣，關閉排氣活門，然後烘焙，待麵糰轉色，立即開啟排氣活門。

1. Knead the dough and add the salt 3 minutes before kneading ends.
2. Add the horseradish and flaxseed after the kneading process is done, in slow speed.
3. After the bulk time, flatten the dough to about 2 cm thick and let it chill in the fridge for about 1 hour, covered with plastic.
4. Cut the dough slab into squares of about 5 cm and place on a baking tray.
5. Once proofed, bake with steam and closed exhaust. Open the exhaust once coloring starts.

建議 SUGGESTIONS

融合了辣根和亞麻籽於粗麥粉麵包，它的味道獨特兼是用餐時的最佳拍擋。嘗試與烤肉享用或配其他愛心美食，均有意想不到的效果。

關於辣根的製法，可用新鮮辣根的莖部，去皮和切成小丁粒，然後放進攪拌機，加少許冰水、鹽和1茶匙醋，攪拌成幼滑質感。

The blend of horseradish and flaxseed into semolina bread is unique and a great match to your dining table. Think about serving it with any roasted meat and other hearty meals.

For the horseradish: Find fresh horseradish roots, peel them and cut them into dice. In a food processor, add a little ice water, salt and a table spoon of vinegar. Blend it until it has the texture of a horseradish cream.

巧味包 Almost untouchable

烘焙術語詞彙 GLOSSARY

刀片 **BLADE**	任何一款鬚刨剃刀、小刀或特別用於麵包刻紋的刀片。 Either razor blade, knife or special blade used to score the bread.
水泡 **BLISTER**	烘焙時，形成於麵包外層的小氣泡。 Small bubbles forming on the crust of the bread while baking.
主體 **BODY**	用於確定麵糰的質地堅硬、柔軟或軟硬適中。 Used to define the texture of dough, either firm or soft or batard (neither firm nor soft).
初次發酵／初步發酵 **BULK PROOF / BULK FERMENTATION**	混合和搓揉麵糰後，第一次發酵的時間。 The first proofing time after the mixing and kneading of dough.
燒焦／過度烘焙 **BURNED**	麵包在烘焙底層石板的溫度太高，令麵包兩邊的底部過度烘焙。 Said of bread that is over baked underneath, with a too high temperature of the stone bed.
燒焦麵糰 **BURNT DOUGH**	過度攪拌麵糰，超過麵糰形成筋性的極限。 Over mixed dough, passed the peak of gluten formation.
布 **CLOTH**	一塊用於蓋住已造型麵糰以備發酵的布。 Piece of cloth used to ferment the shaped loaves.
倒塌／瀉下 **COLLAPSE**	麵糰因混合或造型缺乏韌度而下塌。 Dough which collapses after mixing or shaping due to a lack of strength.
覆蓋／蓋面 **COVER**	覆蓋麵糰表面，防止風乾。 To cover dough to avoid drying out.
出體 **DIVIDING**	麵糰發酵後分成小糰，行內術語是出體。 To divide dough after the bulk fermentation.
降溫 **FALLING**	泛指焗爐的溫度由開始烘焙時而降低。 Said of an oven with temperature decreasing from the start of baking.
最後發酵 **FINAL PROOF**	成形後入爐前的最後發酵。 The last proof after shaping and before baking.
平坦 **FLAT**	泛指麵糰因太軟或麵粉太弱而缺乏力量。 Said of a loaf that lack of strength due to too soft dough or weak flour.
灑粉 **FLOURED**	在工作枱、發酵布或麵糰上灑麵粉。 To dust flour on a working space, a fermentation cloth or a dough.
覆摺 **FOLDING**	麵糰覆摺為了額外的二氧化碳空間，給予麵糰更強韌度。 To fold dough in order to extract the carbonic cells to give more strength to the dough.
搓揉 **KNEADING**	伸展或拉扯麵糰以生產麵筋網絡。 To stretch and pull a dough to create the gluten network.

混合 **MIXING**	在搓揉過程前，將不同的材料形成一團。 To homogenize the ingredients to form dough before the kneading process.
恢復／回復 **REFRESH**	加麵粉和清水，特別是酸酵麵種，目的是進一步繼續發酵。 To add flour and water, especially to sourdough, to further continue the fermentation.
鬆弛／鬆身 **RESTING**	讓已分體的麵糰在造型前鬆身。 To allow the divided pieces of dough to rest before shaping.
升起 **RISE**	指麵糰或麵包膨脹。 Said of dough or loaf when it is expanding.
捲 **ROLL**	造圓形麵糰或麵包的動作。 The action of shaping a round loaf.
刻紋／劃紋 **SCORING**	烘焙前用不同刀片在麵包上刻紋。 Using different blades to score the breads before baking.
造型 **SHAPING**	醒發麵糰之時按照指定形態造型。 To shape a piece of dough into the required shape while extracting gases.
脫皮 **SKIN**	麵糰沒有蓋面，而形成皮層。 A skin formed following the lack of cover on dough.
磚爐 **SOLE**	焗爐的底部一般附有真石頭，時至今日，以由合成石頭取代真石頭。 The bottom of the oven usually made of real stone, nowadays replaced by stone composite.
放入蒸氣 **STEAM**	在焗爐噴出清水，令麵糰表面濕潤，進一步膨脹。 Water injected in the oven to wet the outer layer of the bread in order to allow expansion.
力度／韌度 **STRENGTH**	麵糰或麵包的相關柔韌度。 Said of the relative power of dough or loaves.
收緊麵糰 **TIGHT DOUGH**	當混合時加入麵粉，令麵糰堅硬。 Add flour during the mixing to make the dough firmer.
收緊造型 **TIGHT SHAPING**	麵糰造型時多加點力度，使最後的製品有緊密質地。 To shape a loaf with more strength relative to the texture of dough and desired end product.
水冷卻器 **WATER COOLER**	一具儀器，用作冷卻過高溫度。 Piece of equipment used to cool water where temperature is too high.
接合處 **WELDING**	麵糰的收口／接合位，常於已造型麵糰的底部。 The closing point of a loaf that is usually placed underneath the shaped dough.
年輕麵糰 **YOUNG DOUGH**	麵糰鬆弛不足，在下一步驟繼續。 Said of dough that is not rested enough to carry on to the next step.

烘焙術語詞彙 GLOSSARY

ACKNOWLEDGMENT 鳴謝

我衷心感謝及向香港最優秀的麵包師楊嘉明和 Bread Elements and Pastry Elements 團隊表示極度敬意。在製作新書期間，我得到 Celene Loo, Xavier Honorin and Vincent Thierry 等賜序、建議和支援，銘記深恩。

不過，最大支持莫如是我的摯愛家人如太太 Vianna, 兒子 Clement；爸爸、媽媽、弟弟 Sébastien 和弟婦 Laetitia、岳母麗英和大舅健成，無言感激。

My heartfelt thank you and most respected appreciation to Mark Yeung: Hong Kong's finest baker as well as the entire Bread Elements and Pastry Elements team. Also, a big thank you to Celene Loo, Xavier Honorin and Vincent Thierry for their great help and support.

But most of all and for the greatest support, all my love goes to Vianna, my son Clement, Papa, Maman, Sébastien and Laetitia, Lai Ying and New, Thank you.

熱石上的麵包 LA BOULANGERIE

編著	Author
閔言樂	Grégoire Michaud
編輯	Editor
郭麗眉	Cecilia Kwok
翻譯	Translator
辜安妮	Annie Ko
攝影	Photographer
幸浩生	Johnny Han
封面拍攝	Cover Photographer
	Josephine Rozman
美術設計	Designer
梁施敏	Mandi Leung
出版者	Publisher
飲食天地出版社	Food Paradise Publishing Co.
香港鰂魚涌英皇道1065號	Room 1305, Eastern Centre, 1065 King's Road,
東達中心1305室	Quarry Bay, Hong Kong.
電話	Tel: 2564 7511
傳真	Fax: 2565 5539
電郵	Email: info@wanlibk.com
網址	Web Site: http://www.wanlibk.com
	http://www.facebook.com/wanlibk
發行者	Distributor
香港聯合書刊物流有限公司	SUP Publishing Logistics (HK) Ltd.
香港新界大埔汀麗路36號	3/F., C&C Building, 36 Ting Lai Road,
中華商務印刷大廈3字樓	Tai Po, N.T., Hong Kong
電話	Tel: 2150 2100
傳真	Fax: 2407 3062
電郵	Email: info@suplogistics.com.hk
承印者	Printer
美雅印刷製本有限公司	Elegance Printing & Book Binding Co Ltd.
出版日期	Publishing Date
二零一六年一月第一次印刷	First print in January 2016
版權所有・不准翻印	All rights reserved.

萬里機構

萬里 Facebook

myCOOKey.com